Genebank Standards
for Plant Genetic Resources
for Food and Agriculture

Food and Agriculture Organization of the United Nations
Rome, 2016

Citation: FAO. 2014. *Genebank Standards for Plant Genetic Resources for Food and Agriculture*. Rev. ed. Rome.

Species featured on cover
(from left to right, beginning
with the top row):

Triticum spp.
Capsicum annuum
Anacardium occidentale
Carica papaya
Zea mays
Oryza sativa
Punica granatum
Colocasia esculenta
Phaseolus vulgaris
Araucaria angustifolia
Chenopodium quinoa
Cucurbita maxima
Brachychiton populneus
Phaseolus vulgaris

CONTENTS

Acknowledgements

The preparation and publication of the *Genebank Standards for Plant Genetic Resources for Food and Agriculture* has been possible thanks to the contribution of many individuals. The process has involved contributions from national focal points for plant genetic resources for food and agriculture as well as scientists from national and international organizations. FAO takes this opportunity to sincerely thank them for their time, commitment and expertise.

The Genebank Standards were prepared by FAO's Plant Production and Protection Division under the supervision of Kakoli Ghosh. During its preparation, the FAO team – Kakoli Ghosh, Arshiya Noorani and Chikelu Mba – worked very closely with Ehsan Dulloo, Imke Thormann and Jan Engels from Bioversity International who deserve a special mention. Many thanks also to: Jane Toll from the Global Crop Diversity Trust; Patricia Berjak and Norman Pammenter from the University of KwaZulu-Natal for their excellent contributions. Many FAO staff provided contributions including: Linn Borgen-Nilsen, Stefano Diulgheroff, Alison Hodder, Dan Leskien, NeBambi Lutaladio, Dave Nowell, Michela Paganini and Álvaro Toledo.

We would like to acknowledge and thank the scientists who reviewed the *Genebank Standards:*

Ananda Aguiar, Adriana Alercia, Nadiya AlSaadi, Ahmed Amri, Catalina Anderson, Miriam Andonie, Åsmund Asdal, Sarah Ashmore, Araceli Barceló, Maria Bassols, M. Elena González Benito, Erica E. Benson, Benoit Bizimungu, Peter Bretting, Zofia Bulinska, Marilia Burle, Patricia Bustamante, Emilia Caboni, Lamis Chalak, Rekha Chaudhury, Xiaoling Chen, Andrea M. Clausen, Carmine Damiano, Hadyatou Dantsey-Barry, Maria Teresa Merino De Hart, Axel Diederichsen, Carmen del Río, Ariana Digilio, Sally Dillon, Andreas W. Ebert, David Ellis, Richard Ellis, Florent Engelmann, Epp Espenberg, Francisco Ricardo Ferreira, Brad Fraleigh, R. Jean Gapusi, Massimo Gardiman, Tatjana Gavrilenko, Daniela Giovannini, Agnès Grapin, Badara Gueye, Eva Hain, Magda-Viola Hanke, Jean Hanson, Keith Harding, Siegfried Harrer, Ir Haryono, Fiona R. Hay, Monika Höfer, Kim Ethel Hummer, Salma Idris, Brian M. Irish, Joseph Kalders, Joachim Keller, Maurizio Lambardi, Ulrike Lohwasser, Judy Loo, Xinxiong Lu, Carmen Martín, Rusudan Mdivani, Carlos Miranda, Javad Mozafari, Gregorio Muñoz, Godfrey Mwila, Fawzy Nawar, Normah M. Noor, Dorota Nowosielska, Anna Nukari , Sushil Pandey, Maria Papaefthimiou, Wiesław Podyma, Lerotholi Qhobela, Robin Probert, Alain Ramanantsoanirina, Morten Rasmussen, B.M.C. Reddy, Bob Redden, Barbara M. Reed, Harriet Falck Rehnm, Ken Richards,

Maria Victoria Rivero, Jonathan Robinson, Manuel Sigüeñas Saavedra, Izulmé Rita Santos, Viswambharan Sarasan, Sarah Sensen, Fabiano Soares, Artem Sorokin, Chisato Takashina, Ayfer Tan, Mary Taylor, Mohammed Tazi, Bradley J. Till, Roberto Tuberosa, Rishi Kumar Tyagi, Theo van Hintum, Nguyen Van Kien, Bert Visser, Juan Fajardo Vizcayno, Christina Walters, Wei Wei, Fumiko Yagihashi, and Francis Zee.

Special thanks to Petra Staberg and Pietro Bartoleschi's team for the design and layout of the publication. Thanks are also due to Munnavara Khamidova, Sitora Khakimova, Diana Gutiérrez Méndez, and Suzanne Redfern for their contribution.

There are certainly several others who deserve a mention. Our apologies and thanks are conveyed to all those persons who may have provided assistance for the preparation of the *Genebank Standards* and whose names have been inadvertently omitted.

Foreword

Plant genetic resources are a strategic resource at the heart of sustainable crop production. Their efficient conservation and use is critical to safeguard food and nutrition security, now and in the future. Meeting this challenge will require a continued stream of improved crops and varieties adapted to particular agro-ecosystem conditions. The loss of genetic diversity reduces the options for sustainably managing resilient agriculture, in the face of adverse environments, and rapidly fluctuating meteorological conditions.

Well-managed genebanks both safeguard genetic diversity and make it available to breeders. The *Genebank Standards for Plant Genetic Resources for Food and Agriculture*, prepared under the guidance of the FAO Commission on Genetic Resources for Food and Agriculture, and endorsed at its Fourteenth Regular Session in 2013, lay down the procedures that need to be followed for conservation of plant genetic resources. The Commission recognizes them as being of universal value in germplasm conservation throughout the world.

The voluntary *Standards* cover both seeds in genebanks and vegetatively propagated planting material, including in the field genebanks. They set the benchmark for current scientific and technical best practices, and reflect the key international policy instruments for the conservation and use of plant genetic resources. They are an important tool in implementing the *International Treaty on Plant Genetic Resources for Food and Agriculture*, and a supporting component of the *Second Global Plan of Action for Plant Genetic Resources for Food and Agriculture*. The world's 7.5 million genebank accessions are largely of the crops on which humans and livestock most rely for food and feed, including important wild relatives and landraces, but others are of crops of local importance and underutilized species.

The *Standards* encourage active genebank management, and provide for a set of complementary approaches. They will help genebank managers strike a balance between scientific objectives, resources available, and the objective conditions under which they work, recognizing that the world's over 1 750 genebanks differ greatly in the size of their collections and the human and financial resources at their disposal. The challenge that many developing countries face in ensuring secure long-term conservation, in the face of limited capacities and inadequate infrastructure, makes this a challenging task.

The value of conserving crop genetic resources is realized only through their effective use. This requires strong linkages along the chain from *in situ* resource conservation and collection, through storage in genebanks, through research and breeding, to farmers and their communities, and ultimately consumers. Genebank curators, breeders, and national programmes must work hand in hand to ensure the efficient and sustainable conservation of the plant genetic resources for food and agriculture on which humanity depends. I call for adequate provision to be made at national and regional level, so that these crucial international Standards can fulfil their objective of underwriting food security.

Ren Wang
Assistant Director-General
Agriculture and Consumer Protection Department

Preface

Genebanks play a key role in the conservation, availability and use of a wide range of plant genetic diversity for crop improvement for food and nutrition security. They help bridge the past and the future by ensuring the continued availability of genetic resources for research, breeding and improved seed delivery for a sustainable and resilient agricultural system. An efficient management of genebanks through application of standards and procedures is essential for the conservation and sustainable use of plant genetic resources.

The *Genebank Standards for Plant Genetic Resources for Food and Agriculture* (Genebank Standards) provide international standards for ex situ conservation in seed banks, field genebanks and for in vitro and cryopreservation. The Seeds and Plant Genetic Resources Team prepared the Standards under the guidance of the Commission on Genetic Resources for Food and Agriculture. During the preparatorwy phase, standards for orthodox seeds were updated and others developed for field genebanks and for *in vitro* and cryopreservation in consultation with the CGIAR, in particular Bioversity International. Genebank managers, relevant academic and research institutions, national focal points for plant genetic resources for food and agriculture have been instrumental in providing valuable feedback. This was also true for the Secretariats of the International Treaty on Plant Genetic Resources for Food and Agriculture and the International Plant Protection Convention. At its Fourteenth Session in 2013, the Commission endorsed the *Genebank Standards* and urged their universal adoption.

The aim of the Genebank Standards is the conservation of plant genetic resources under conditions that meet recognized and appropriate standards based on current and available technological and scientific knowledge. All the standards are founded on underlying principles that are common to all the different types of genebanks. They also take into account the changes in seed management and techniques due to advances in molecular biology, and bioinformatics. They incorporate the developments in the field of documentation and information systems that are increasingly becoming central to improving genebank management and optimization of resources. A narrative describing the context, technical aspects, contingencies and selected references on technical manuals and protocols as appropriate, supports each standard in the document.

The Genebank Standards are generic enough to be applicable to all genebanks and should be used in conjunction with species-specific information. These is especially true for plants producing non-orthodox seeds and/or are vegetatively propagated as it is difficult to establish specific standards that are valid for all those species given their different seed storage behaviours, life forms and life cycles. The standards are nonbinding and voluntary and stress the importance of securing and sharing material along with related documentation in line with national and international regulations. It will be useful for the standards to be reviewed periodically taking into account the changing policy and technical landscapes.

Conserving and increasing the sustainable use of plant genetic resources is a necessary for achieving food security and addressing nutritional requirements of present and future generations. Therefore, it is vital to conserve the diversity of plant genetic resources so that it is available to the global community. However, genebank maintenance can be expensive. Many scientific advances, such as cryopreservation, come at a cost, especially when used for large-scale testing. Maintenance of field genebanks is equally demanding in terms of labour and cost. Therefore, the emphasis should be on proactive management of genebanks by adopting a complementary approach, and striking an optimal balance between scientific considerations, available personnel, infrastructural and financial resources under prevailing conditions. In many countries, the availability of trained personnel and adequate resources to maintain genebank collections in a sustainable manner remains a challenge. Long-term partnerships at national, regional and global levels together with resources for capacity development will be necessary to apply the standards.

Chapter **1**

Introduction

Seed and Plant Genetic Resources Team

Genebanks around the world hold collections of a broad range of plant genetic resources, with the overall aim of long-term conservation and accessibility of plant germplasm to plant breeders, researchers and other users. Plant genetic resources are the raw materials utilized in crop improvement and their conservation and use is critical to global food and nutrition security. Sustainable conservation of these plant genetic resources depends on effective and efficient management of genebanks through the application of standards and procedures that ensure the continued survival and availability of plant genetic resources.

The *Genebank Standards for Plant Genetic Resources for Food and Agriculture* arises from the revision of the FAO/IPGRI *Genebank Standards*, published in 1994. The revision was undertaken at the request of the Commission on Genetic Resources for Food and Agriculture (CGRFA) in light of changes in the global policy landscape and advances in science and technology. The main policy developments that impact the conservation of plant genetic resources for food and agriculture (PGRFA) in genebanks lie within the context of availability and distribution of germplasm arising from the adoption of various international instruments. These include the Convention on Biological Diversity (CBD), the International Treaty on Plant Genetic Resources (ITPGRFA), the International Plant Protection Convention (IPPC) and the WTO Sanitary and Phytosanitary Agreement (WTO/SPS). In 2010, the CBD adopted the Nagoya Protocol on Access to Genetic Resources and Equitable Sharing of Benefits Arising from their Utilization, which has potential for impact upon germplasm exchange. On the scientific front, advances in seed storage technology, biotechnology and information and communication technology have added new dimensions to plant germplasm conservation.

The *Genebank Standards for Plant Genetic Resources for Food and Agriculture* is intended as a guideline for genebanks conserving plant collections (seeds, live plants and explants). They were developed based on a series of consultations with a large number of experts in seed conservation, cryopreservation, *in vitro* conservation and field genebanks worldwide. The standards are voluntary and nonbinding and have not been developed through standard-setting procedure. They should be viewed more as targets for developing efficient, effective and rational *ex situ* conservation in genebanks that provides optimal maintenance of seed viability and genetic integrity, thereby ensuring access to, and use of, high quality seeds of conserved plant genetic resources.

It is important that these Genebank Standards are not used uncritically as there are continuous technological advances in conservation methods, much of it species-specific, as well as in the context of the purpose and period of germplasm conservation and use. It is recommended that the Genebank Standards should be used in conjunction with other reference sources, particularly with regards to species-specific information. This is especially true for plants producing non-orthodox seeds and /or are vegetatively propagated, of which there exist different seed storage behaviours, life forms (herbs, shrubs, trees, lianas/vines) and life cycles (annual, biennial, perennial) for which it is difficult to establish specific standards that are valid for all species.

This document is divided into two parts. The first part describes underlying principles that underpin the Genebank Standards and provide the overarching framework for effective and efficient management of genebanks. The key principles at the core of genebank operation are the preservation of germplasm identity, maintenance of viability and genetic integrity, and the promotion of access. This includes associated information to facilitate use of stored plant material in accordance with relevant national and international regulatory instruments. The underlying principles are common to all the different types of genebanks.

The second part provides the detailed standards for three types of genebanks namely: seed banks, field genebanks and *in vitro*/cryopreservation genebanks. The standards cover all the major operations carried out in genebanks and a selective list of references is provided for all standards. While key technical information is provided for the standards, it is important to note that appropriate technical manuals should be consulted for procedures and protocols. The seed bank standards (Chapter 4) deals with the conservation of the desiccation-tolerant orthodox seeds, i.e. can be dehydrated to low water content and are responsive to low temperatures. Lowering moisture and temperature decreases the rate of metabolic processes, thus increasing seed longevity. Examples of orthodox-seeded plants include maize (*Zea mays* L.), wheat (*Triticum* spp.), rice (*Oryza* spp.), chickpea (*Cicer arietinum*), cotton (*Gossypium* spp.) and sunflower (*Helianthus annuus*).

Standards for field genebank and *in vitro* conservation/cryopreservation genebanks aimed at the conservation of plants that produce non-orthodox seeds, also known as recalcitrant or intermediate seeds, and/or are propagated vegetatively, are provided in Chapters 5 and 6 respectively. Such plants cannot be conserved in the same way as orthodox seeds, i.e., at low temperature and humidity and require other methods of *ex situ* conservation.

Field genebanking is the most commonly used method for non-orthodox seed producing plants. It is also used for plants that produce very few seeds, are vegetatively propagated and/or plants that require a long life cycle to generate breeding and/or planting materials. Although the term 'field genebank' is used, the method also includes the maintenance of live plants in pots or trays in greenhouses or shade houses. Technical guidelines and training manuals are available for the management of germplasm collections held in field genebanks (e.g. Bioversity International *et al.* 2011; Reed *et al.*, 2004; Said Saad and Rao, 2001; Engelmann, 1999; Engelmann and Takagi, 2000; Geburek and Turok, 2005).

The conservation of plant germplasm *in vitro* and cryopreservation can either be conserved through slow growth (*in vitro*) for short/medium-term storage, or cryopreservation for long-term conservation. The former method involves cultures (especially shoot tips, meristems, somatic embryos, cell suspension or embryogenic callus) being maintained under growth-limiting conditions on artificial culture media. The growth rate of the cultures can be limited by various methods, including temperature reduction, lowering of light intensity, or manipulation of the culture medium by adding osmotic agents or growth retardants (Engelmann, 1999).

Cryopreservation is the storage of biological materials (seeds, plant embryos, shoot tips/meristems, and/or pollen) at ultra-low temperatures, usually that of liquid nitrogen (LN) at −196 °C (Engelmann and Takagi, 2000; Reed, 2010). Under these conditions, biochemical and most physical processes are halted and materials can be conserved over the long term. These modes of conservation constitute a complementary approach to other modes and are necessary for safe, efficient and cost effective conservation (Reed, 2010). For example, cryopreserved lines can be maintained as a backup for field collections, as reference collections for available genetic diversity of a population, and as a source for new alleles in the future.

The following standards are provided for the respective type of genebank:

- **Genebank Standards for Orthodox Seeds:** acquisition of germplasm, seed drying and storage, viability monitoring, regeneration, characterization, evaluation, documentation, distribution, safety duplication and security/personnel.

- **Field Genebank Standards:** choice of location, acquisition of germplasm, establishment of field collections, field management, regeneration and propagation, characterization, evaluation, documentation, distribution, security and safety duplication.

- **Genebank Standards For *In Vitro* Culture and Cryopreservation:** acquisition of germplasm, testing for non-orthodox behaviour and assessment of water content, vigour and viability, hydrated storage for recalcitrant seeds, *in vitro* culture and slow-growth storage, cryopreservation, documentation, distribution and exchange, security and safety duplication.

SELECTED REFERENCES

Bioversity International/Food and Fertilizer Technology Center/TARI-COA (Taiwan Agricultural Research Institute-Council of Agriculture). 2011. *A training module for the international course on the management and utilisation of field genebanks and* in vitro *collections.* Fengshan, Taiwan, TARI.

Engelmann, F., ed. 1999. *Management of field and in vitro germplasm collections.* Proceedings of a Consultation Meeting, 15–20 January 1996. Cali, Colombia, CIAT, and Rome, IPGRI.

Engelmann, F. & Takagi, H., eds. 2000. *Cryopreservation of tropical plant germplasm. Current research progress and application.* Tsukuba, Japan, Japan International Research Center for Agricultural Sciences, and Rome, IPGRI.

Geburek, T. & Turok, J., eds. 2005. *Conservation and management of forest genetic resources in Europe.* Zvolen, Slovakia, Arbora Publishers.

Reed, B.M. 2010. *Plant cryopreservation. A practical guide.* New York, USA, Springer.

Reed, B.M., Engelmann, F., Dulloo, M.E. & Engels, J.M.M. 2004. *Technical guidelines for the management of field and* in vitro *germplasm collections.* Handbooks for Genebanks No. 7. Rome, IPGRI.

Said Saad, M. & Rao, V.R., eds. 2001. *Establishment and management of field genebank training manual.* Serdang, Malaysia, IPGRI-APO.

SGRP-CGIAR (System-wide Genetic Resources Programme of the Consultative Group on International Agricultural Research). Crop Genebank Knowledge Base (available at: http://cropgenebank. sgrp.cgiar.org).

Chapter 2

Underlying principles

Genebanks across the globe share many of the same basic goals, but their missions, resources, and the systems they operate within, often differ. As a result, curators have to optimize their own genebank systems and this requires management solutions that may differ substantially across institutions while achieving the same objectives. Underlying principles explain why and for what purpose plant genetic resources are being conserved. These principles provide the basis for establishing the norms and standards essential for the smooth operation of a genebank. The major underlying principles for conservation are described in the section below.

Identity of accessions

Care should be taken to ensure that the identity of seed sample accessions conserved in genebanks is maintained throughout the various processes, beginning with acquisition through to storage and distribution. Proper identification of seed samples conserved in genebanks requires careful documentation of data and information about the material. This begins with recording passport data and collecting donor information if applicable. Where possible, such information should also be recorded for older collections in genebanks for which passport data was not previously recorded or is incomplete. Herbarium voucher specimen and seed reference collections often play an important role in the correct identification of seed samples. Identification of accessions in the field is especially important since inadequate labelling can lead to much genetic erosion. Field labelling needs to be complemented with field layout plans, which should be properly documented in order to ensure proper identification of accessions in field genebanks.

Field labels are prone to loss due to various external factors, e.g. bad weather conditions. Modern techniques, such as accession labels with printed barcodes, Radio- Frequency Identification (RFID) tags and molecular markers, can greatly facilitate the management of germplasm by reducing the possibility of error, further ensuring the identity of accessions.

Maintenance of viability

Maintaining viability, genetic integrity and quality of seed samples in genebanks and making them available for use is the ultimate aim of genebank management. Therefore, it is critically important that all genebank processes adhere to the standards necessary to ensure that acceptable levels of viability are maintained. To achieve this, particular attention needs to be paid to standards on germplasm acquisition, processing and storage. For recalcitrant and other non-orthodox seed types, this is assessed by visual inspection for lack of damage, and by rate and totality of germination. However, the occurrence of macroscopically undetectable fungi and bacteria within the seeds may compromise seed quality. In seed genebanks generally, seed samples accepted into the genebank at the point of acquisition should have high viability and as far as possible meet the standards for acquisition of germplasm. Collecting the seeds as close as possible to maturation but prior to natural dispersal, avoiding collection of dispersed seeds from the ground or those that are soiled and may have saprophytic or pathogenic fungi/bacteria, will help ensure the highest physiological seed quality. Genebanks, to the extent possible, should ensure that collected germplasm is genetically representative of the original population as well as take into account the number of live propagules, such that sample quality is not compromised. A monitoring system should be in place to check the viability status of stored samples at appropriate intervals depending on expected seed longevity. Frequency of regeneration can be reduced if correct attention is paid to post-harvest handling, drying and storage. In the context of field genebanks, 'propagability' (i.e. the quality and state of being propagable) is more relevant than 'viability', which specifically relates to the capacity of seeds to germinate and produce a plantlet. Field genebanks are vulnerable to the impacts of environmental factors such as weather conditions, incidence of pests, etc. and the extent of such impacts will differ according to different species type and growth cycles e.g. annual, biennial or perennial. An additional factor in the case of species with seeds of unknown post-storage behaviour (i.e. whether recalcitrant, otherwise non-orthodox or orthodox) is the *a priori* requirement of ascertaining seed responses (generally to slow dehydration) before any germplasm storage strategy can be put in place.

Maintenance of genetic integrity

The need to maintain genetic integrity is closely related to maintenance of the viability and diversity of the original collected sample. All genebank processes, starting from collection and acquisition, through to storage, regeneration and distribution, are important for the maintenance of genetic integrity. Ensuring that viability is maintained according to the standards contributes to the maintenance of genetic integrity. Various molecular techniques, including surveys of possible epigenetic changes that may or may not be reversible, are needed to assess whether genomic stability has been maintained, particularly when samples are retrieved from cryostorage. In plants requiring long intervals from sowing to reproductive maturity, seed regeneration in the field would be highly impractical. Re-sampling the original population should be undertaken when there are signs of declining vigour and viability. Maintenance of genetic integrity is equally as important for germplasm conserved *in vitro*, especially in view of the risk of somaclonal variation. This is the main reason for avoiding indirect somatic embryogenesis (i.e. via a callus stage) to generate forms of germplasm to be conserved whenever possible. Adequately representative seed samples of good quality and sufficient quantity should be obtained during acquisition as far as achievable. However, it is recognized that when the objective is to collect particular traits, then the sample may not be representative of the original population. To minimize genetic erosion it is important to follow recommended protocols[1] for regenerating seed accessions with as few regeneration cycles as possible, sufficiently large effective population sizes, balanced sampling, as well as pollination control. Special mention is made here of the importance of safety duplication to respond to risks that can occur in genebank facilities.

Maintenance of germplasm health

Genebanks should strive to ensure that the seeds they are conserving and distributing are free from seed-borne diseases and regulated pests (bacteria, virus, fungi and insects). External surfaces can usually be eliminated effectively using surface disinfection procedures. Genebanks often do not have the capacity or necessary resources to

1 **Dulloo, M.E., Hanson, J., Jorge, M.A. & Thormann, I.** 2008. Regeneration guidelines: general guiding principles. In: M.E. Dulloo, I. Thormann, M.A. Jorge & J. Hanson, eds. *Crop specific regeneration guidelines*. CGIAR System-wide Genetic Resource Programme (SGRP), Rome, Italy. 6 pp. Also see: http://cropgenebank.sgrp.cgiar.org.

test whether samples collected or otherwise acquired, and samples harvested from regeneration/multiplication plots, are free from seed-borne diseases and pests. This is particularly the case with germplasm received from third parties. The problems are exacerbated for the conservation of recalcitrant-seeded species. Internally-borne contaminants are revealed only when recalcitrant seeds are maintained in short- to medium-term hydrated storage, or when seed-derived explants are placed in tissue culture. The presently unsatisfactory solution is to discard any contaminated seed/explant, as it is the only way to ensure uncontaminated germplasm. Thus, it is important that relevant import and phytosanitary certificates accompany seed material when exchange of germplasm takes place to ensure the health status of samples received. Some infected/infested samples may be easily cleaned, while others may require more elaborate methods for cleaning.

Physical security of collections

An underlying principle of germplasm conservation is that the physical structures of the genebank facilities in which germplasm are conserved are of adequate standard to secure the materials from any external factors, including natural disasters and human-caused damage. Adequate security systems are required to ensure that genebank cooling equipment, as well as backup generators and equipment to control power outages, are in good running condition and monitoring devices are available to track essential parameters over time. As cryogenic storage requires LN, supplies of this cryogen must always be available. Furthermore, it is vital that levels of LN are maintained, whether filling/top-up of the special storage vats or LN freezers used is done manually or automatically. Another important security issue for genebanks is to ensure materials are safely duplicated in other locations so that if a collection suffers loss, for any reason, material can be restored from duplicated sets.

Availability and use of germplasm

Conserved material must be available for current and future use. It is, therefore, important that all processes in genebank operations and management contribute to this goal. There will be a need to maintain sufficient quantities of seed and related information on the accessions. Although there are a few individuals of accessions in field genebanks, and thus a limited capacity for distributing to users, the genebank should have a strategy in place to multiply quickly any germplasm for distribution.

Availability of information

In order to ensure communication of information and accountability, essential, detailed, accurate, and up-to-date information – including historical as well as current information, especially in relation to the management of individual accessions subsequent to their acquisition – should be recorded in electronic databases. Access, availability and sharing of this information should be treated with high priority, as it leads to better and more rational conservation. Search-query interactive databases containing phenotypic evaluation data can assist germplasm clients in the targeting of germplasm requests and, in turn, feedback of further evaluation data adds to the value and utility of the collection. If information on the conserved germplasm is made easily available and accessible, it will enhance germplasm use. Further, this will help the genebank curators to better plan their multiplication and regeneration activities in order to keep adequate stocks of their accessions. For such genebank-based information systems, a search-query interactive database is recommended. The Seed Information Database (SID)[2] of the Millennium Seed Bank (MSB), Kew, provides a good example of the value of this type of database. BRAHMS (Botanical Research and Herbarium Management System)[3] is a system developed for the purposes of curation and germplasm data management, while EURISCO[4] is a web-based catalogue that provides information about European *ex situ* plant collections.

Proactive management of genebanks

Sustainable and effective conservation of genetic resources depends on active management of conserved germplasm. Proactive management is critical for ensuring that germplasm is efficiently conserved and made timely available and in adequate quantity for further use by plant breeders, farmers, researchers and other users. It emphasizes the importance of securing and sharing material as well as the related information, and sets in place a functional strategy for management of human and financial resources for a rational system. It includes a risk management strategy and encourages collaborations with third parties in providing services to genebanks in the efforts to conserve biodiversity. It should be mentioned that the

---

2   Seed Information Database: http://data.kew.org/sid.

3   BRAHMS: http://dps.plants.ox.ac.uk/bol.

4   EURISCO: http://eurisco.ecpgr.org.

maintenance of field collections is costly and all efforts should be made to develop complementary collections *in vitro* or in cryopreservation. Adherence to the legal and regulatory frameworks at national and international levels, in particular as they relate to access, availability and distribution of materials and plant and seed health is necessary. Where appropriate, a Standard Material Transfer Agreement (SMTA) should be used for crops under the Multilateral System of the ITPGRFA. The IPPC regulations provide the framework for quarantine and health regulations to prevent the introduction and spread of plant pests and diseases. Above all, there is a need for long-term and continuous commitment of the institutions holding genebanks with regard to the availability of human and financial resources.

Furthermore, proactive management encourages application of practical experiences and knowledge to new germplasm in a genebank and seeks to apply the Genebank Standards to the extent possible under locally prevailing conditions. This could sometimes mean that, although a particular standard is not entirely met, precautionary measures are taken to uphold the underlying principles of genebank management.

Chapter 3

Standards – structure and definitions

The Standards, as described in this document, define the level of performance of a routine genebank operation below which there is a high risk of losing genetic integrity (e.g. a probability of 5 percent or more of losing an allele in an accession over the storage period). Each section is divided into:

- *Standards*
- *Context*
- *Technical aspects*
- *Contingencies*
- *Selected references*

The **CONTEXT** *provides the basic necessary information in which the standards apply. It provides a brief description of the routine genebank operation for which the standards are defined and the underlying principles for them. The* **TECHNICAL ASPECTS** *explain technical and scientific principles important to understand and underpin the standards. The* **CONTINGENCIES** *are provisions in case standards cannot be applied e.g. to a given species. It covers exceptions, alternative routes, and risk management options. Selected sources of* **INFORMATION AND REFERENCES** *are provided in all sections.*

Chapter **4**

Genebank standards for orthodox seeds

4.1   Standards for acquisition of germplasm

Standards

4.1.1   All seed samples added to the genebank collection have been acquired legally with relevant technical documentation.

4.1.2   Seed collecting should be made as close as possible to the time of maturation and prior to natural seed dispersal, avoiding potential genetic contamination, to ensure maximum seed quality.

4.1.3   To maximize seed quality, the period between seed collecting and transfer to a controlled drying environ ment should be within 3 to 5 days or as short as possible, bearing in mind that seeds should not be exposed to high temperatures and intense light and that some species may have immature seeds that require time after harvest to achieve embryo maturation.

4.1.4   All seed samples should be accompanied by at least a minimum of associated data as detailed in the FAO/Bioversity multi-crop passport descriptors.

4.1.5   The minimum number of plants from which seeds should be collected is between 30-60 plants, depending on the breeding system of the target species.

Context

Acquisition is the process of collecting or requesting seeds for inclusion in the genebank, together with related information. The material should be legally acquired, be of high seed quality and properly documented.

Acquisition is made in accordance with relevant international and national regulations such as phytosanitary/quarantine laws, ITPGRFA or CBD access regulations, and national laws for genetic resources access. Adherence to Standard 4.1.1 will allow the export of seeds from the origin/donor country and the import into the country of the genebank, and determine the management and distribution regime (e.g. SMTA or Material Transfer Agreements [MTA]).

There is a need to ensure maximum seed quality and avoid conservation of immature seeds and seeds that have been exposed for too long to the elements. The way that seeds are handled after collection and before they are transferred to controlled conditions is critical for seed quality. Unfavourable extreme temperatures and humidity during the post-collecting period and during transport to the genebank could cause rapid loss in viability and reduce longevity during storage. The same applies to post-harvest handling within the genebank. The seed quality and longevity is affected by the conditions experienced prior to storage within the genebank. It is recommended that a germination test be conducted immediately after processing and before pre-storage as a way to determine the quality of the seed collected.

During the acquisition phase, it is important to ensure that passport data for each accession is as complete as possible and fully documented, especially georeferenced data that can help to locate collection sites. Passport data are crucial in identifying and classifying the accession and will function as entry points in selecting and using the accession.

Technical aspects

Access to PGRFA, in the multilateral system of the International Treaty, has to be accompanied with a SMTA. The acquirers should comply with the relevant provisions of the ITPGRFA or the CBD and a MTA should be signed by the authorized person in the country of collecting, according to national laws for access to genetic resources of the country where the collecting will take place (ENSCONET, 2009). In addition, when required by the providing country, the access should be subject to the prior informed consent of the country. Phytosanitary regulations and any other import requirements must be sought from the relevant national authority of the receiving country.

Seeds that are freshly harvested from the field may have high water content and need to be ventilated to prevent fermentation. They should be placed into suitable containers that allow for good air circulation, and that ensure the contents do not become moist through inadequate air exchange and are neither mixed nor damaged during collecting and transport. Monitoring the temperature and relative humidity (RH) to ensure that seeds are not exposed to conditions above 30 °C or 85 percent RH after collecting and transport, as well as during post-harvest processing will help to maintain seed quality. If fully mature seeds need to be processed and dried in the field, technical recommendations for the particular or similar species should be applied to reduce the risk of deterioration.

Appropriate collecting forms should be used to capture collection data. These forms should include information such as the initial taxonomic classification of the sample, the global positioning system (GPS) coordinates of the collecting site, a description of the habitat of the collected plants, the number of plants sampled and other relevant data that are important for proper conservation. If possible, the FAO/Bioversity multi-crop passport descriptors should be used (Alercia *et al.*, 2012). Very useful additional information, such as cultural practices, previous generations of seed history and origin, uses etc, can be obtained by farmer interviews when seed is collected from farmer fields/stores. During collecting, the collector should also be sensitive to the depletion of the natural population targeted for collecting. The European Native Seed Conservation Network (ENSCONET) collecting manual recommends that the collecting must not overdraw 20 percent of the total seeds available in a population (ENSCONET, 2009). It may also be useful to repeat sampling from a particular site to maximize capture of genetic variability that may be present at various points in time.

The collection sample should be sufficient to include at least one copy of 95 percent of the alleles that occur within the target population with a frequency greater than 0.05 (Marshall and Brown, 1975). A random sample of 59 unrelated gametes is sufficient to achieve this objective and in a species mating complete at random this equates to 30 individuals whereas in a completely selfing species, this target requires 60 individuals (Brown and Hardner, 2000). Thus, the sample size to capture 95 percent of the alleles can vary between 30 and 60 plants depending on the breeding system of the target species. In practice, adequate quantities of seeds should be collected for distribution in order to avoid frequent regeneration. However, we should recognize that this target may not always be met depending on the availability of seeds for collection.

In case of donation of the seeds (from a seed company, research programme or genebank), the taxonomic classification, donor, identification number of the donor, and names in addition to the available passport data should be provided. Adequate information about how the germplasm received was maintained should be sought from the donor, including pedigree or lineage information, as well as chain of custody

information where available. Seeds should be assigned a unique identification number (either temporary or permanent, according to the practice used in the genebank) that accompanies the seeds at all times, and that will link the seeds to the passport data and any other collected information, and guarantee the authenticity of the seed sample. Whenever possible, a herbarium voucher specimen collected from the same population as the seed samples should be taken, and a record should be made of the method and reason for acquisition.

Contingencies

Seeds collected in the field are rarely in such condition (physiological and phytosanitary status) and quantities that long-term conservation is automatically guaranteed. In this case, multiplication in controlled conditions for the specific purpose of long-term conservation is recommended.

When collections contain a significant proportion (>10 percent) of immature seeds or fruits, measures should be taken to encourage post-harvest ripening. This can usually be achieved by holding material in well ventilated, ambient conditions protected from rainfall. Visual improvements in maturity should be monitored and the material should be transferred to controlled drying conditions as soon as the collected seeds are deemed more mature.

Allowances in terms of above standards (e.g. sample size) will have to be made for wild and rare species where seeds might not be available in optimal conditions or quantity.

SELECTED REFERENCES

Alercia, A., Diulgheroff, S. & Mackay, M. 2012. FAO/Bioversity *Multi-Crop Passport Descriptors* (MCPD V.2). Rome, FAO and Bioversity International (available at: http://www.bioversityinternational.org/uploads/tx_news/1526.pdf).

Brown, A.H.D. & Hardner, C.M. 2000. *Sampling the genepools of forest trees for ex situ conservation. In* A. Young, D. Boshier & T. Boyle. *Forest conservation genetics. Principles and practice*, pp.185–196. CSIRO and CABI.

Engels, J.M.M. & Visser, L., eds. 2003. *A guide to effective management of germplasm collections.* Handbooks for Genebanks No. 6. Rome, IPGRI.

ENSCONET (European Native Seed Conservation Network). 2009. *Seed collecting manual for wild species* (available at: http://ensconet.maich.gr/).

Eymann, J., Degreef, J., HŠuser, C., Monje, J.C., Samyn, Y. & VandenSpiegel, D., eds. 2010. *Manual on Field Recording Techniques and Protocols for All Taxa Biodiversity Inventories and Monitoring*, Vol. 8. (available at: http://www.abctaxa.be/volumes/volume-8-manual-atbi).

FAO/IPGRI. 1994. *Genebank standards.* Rome, FAO and IPGRI (available at: ftp://ftp.fao.org/docrep/fao/meeting/015/aj680e.pdf).

Guarino, L., Rao R.V. & Reid, R., eds. 1995. *Collecting plant genetic diversity. Technical guidelines*, Wallingford, UK, CAB International.

Guerrant, E.O., Havens, K. & Maunder, M., eds. 2004. Ex situ *plant conservation: supporting species survival in the wild.* Washington, DC, Island Press.

Lockwood, D.R., Richards, C.M. & Volk, G.M. 2007. Probabilistic models for collecting genetic diversity: comparisons, caveats and limitations. *Crop Science*, 47: 859–866.

Marshall, D.R. & Brown, A.H.D. 1975. Optimum sampling strategies in genetic resources conservation, pp. 53–80. *In* O.H. Frankel & J.G. Hawkes, eds. *Crop genetic resources for today and tomorrow.* Cambridge, UK, Cambridge University Press.

Probert, R., Adams, J., Coneybeer, J., Crawford, A. & Hay, F. 2007. Seed quality for conservation is critically affected by pre-storage factors. *Australian Journal of Botany*, 55: 326–335.

Probert, R.J. 2003. Seed viability under ambient conditions and the importance of drying. *In* R.D. Smith, J.B. Dickie, S.H. Linington, H.W. Pritchard & R.J. Probert, eds. *Seed conservation: turning science into practice*, pp. 337–365. Kew, UK, Royal Botanic Gardens.

Royal Botanic Gardens, Kew. *Millennium Seed Bank technical information sheet 04: post-harvest handling of seed collections.* Kew, UK (available at: http://www.kew.org/msbp/scitech/publications/04-Post%20harvest%20handling.pdf).

SGRP-CGIAR. Crop Genebank Knowledge Base (available at: http://cropgenebank.sgrp.cgiar.org).

Smith, R.D., Dickie, J.B., Linington, S.H., Pritchard, H.W. & Probert, R.J., eds. 2003. *Seed conservation: turning science into practice.* Kew, UK, Royal Botanic Gardens (available at: http://www.kew.org/msbp/scitech/publications/sctsip.htm).

Upadhyaya, H.D. & Gowda, C.L.L. 2009. *Managing and enhancing the use of germplasm –strategies and methodologies.* Technical Manual No. 10. Patancheru, India, International Crops Research Institute for the Semi-Arid Tropics.

4.2 Standards for drying and storage

Standards

4.2.1 All seed samples should be dried to equilibrium in a controlled environment of 5-20 °C and 10-25 percent of relative humidity, depending upon species.

4.2.2 After drying, all seed samples need to be sealed in a suitable airtight container for long-term storage; in some instances where collections that need frequent access to seeds or likely to be depleted well before the predicted time for loss in viability, it is then possible to store seeds in non-airtight containers.

4.2.3 Most-original-seed-samples and safety duplicate seed samples should be stored under long-term conditions at a temperature of –18 ± 3 °C and relative humidity of 15 ± 3 percent.

4.2.4 For medium-term conditions, seed samples should be stored under refrigeration at 5-10 °C and relative humidity of 15 ± 3 percent.

Context

Maintaining seed viability is a critical genebank function that ensures germplasm is available to users and is genetically representative of the population from which it was acquired (i.e. the most-original-sample). A critical objective of seed drying and storage standards is to reduce the frequency of regeneration of the most-original-sample by maximizing seed longevity, thereby reducing the cost of genebanking and the risks of genetic erosion. For this purpose, long-term storage is required for all most-original samples and for safety duplication of the collection (see Standards for safety duplication). In addition storage standards are also required for circumstances where the objective is to store seeds over the medium- or short-term to keep them alive long enough for distribution to users and evaluation of germplasm. In such cases, the standard need not be as stringent as in the case of long-term conservation.

Prior to storage, seed samples need to be dried to appropriate moisture content. A variety of methods can be used for seed drying, the most common being the use of a desiccant or using a dehumidified drying chamber. The methods chosen will depend on the available equipment, number and size of the samples to be dried, local climatic conditions and cost considerations. However, there is a limit to which drying can increase longevity. At a critical moisture level, maximum longevity for the storage temperature is attained and drying below this level does not increase seed longevity further. To realize the full benefit of refrigerated or freezer storage, it is recommended that genebanks dry seeds to the critical moisture level. Various RH-temperature combinations can be used during drying, with faster drying possible at higher temperatures but the potential for physiological aging reduced by lower drying temperatures.

Long-term storage conditions as recommended above are expected to provide high seed quality for long periods, the actual timing is species-specific; medium-term storage conditions are adequate for 30 years and will generally require refrigerated storage. Short-term storage is expected to provide high quality seed for at least eight years and may be accomplished at ambient temperatures (under as cool and stable temperatures as possible but not more than 25 °C) for some longer-lived species if RH is controlled according to Standard 4.2.2. It should be pointed out that the longevity of mature, high quality seeds may vary among species and even among seed lots of the same species (Probert *et al.*, 2009; Nagel and Börner, 2010; Walters *et al.*, 2005). The variation among species and among seed lots of the same species, particularly if seeds are harvested with variable maturity, requires the genebank curator's vigilance to monitor viability (see Standards for viability monitoring).

As seed equilibrium moisture content varies depending on oil content, the best measurement for the drying standard is equilibrium relative humidity (eRH) which is constant depending on the RH and temperature of the drying environment. However, it should be noted that in sealed containers during storage, seed eRH will fall or increase if the storage temperature is lower or higher than the drying temperature.

Technical aspects

Seed longevity is determined by interactions of biological factors intrinsic to the seed and the quality and consistency of the storage environment, namely the storage temperature and the control of seed moisture content (eRH) as well as being species dependent. It is well known that seed longevity increases as the seed moisture content and storage temperature decreases, within limits (Ellis and Roberts, 1980; Harrington, 1972). Studies have demonstrated that drying seed beyond certain critical seed moisture content provides little or no additional benefit to longevity (Ellis *et al.*, 1985; Ellis and Hong, 2006) and may even accelerate seed aging rates (Vertucci and Roos, 1990; Walters, 1998). The storage standards as presented are intended to ensure that seeds are stored at this optimum moisture content. However, it has been shown that lowering the storage temperature increases the optimum seed moisture content level (Walters and Engels, 1998; Ellis and Hong, 2006), which suggests there might be danger of over-drying seeds.

Drying conditions that achieve the critical moisture level at the storage temperature should be determined using water sorption isotherms that show the relationship between the amount of water in the seeds, usually expressed as a percentage of the total seed weight, and their RH. There could be different combinations of RH and drying temperature for given species. Isotherm relationships, predicted based on seed oil content, are available online at the Kew SID website (see references). Genebank operators should clearly understand the relationship between RH and storage temperature to be able to decide about the best combination for their seed drying environment.

As soon as the seeds have reached the desired moisture content, they should be packaged and stored. After drying, seed moisture should be maintained using moisture-proof containers. Different types of containers can be used including glass, tin, plastic containers, and aluminium foils, each with their advantages and disadvantages (Gómez-Campo, 2006). In any case, either glass containers that are sufficiently thick to avoid breakage or laminate packaging with a metal foil layer of adequate thickness will maintain desired moisture levels for up to 40 years, depending on the ambient RH at the genebank's location and the quality of the seal. For example, in Germany the genebank uses laminated aluminium foils that are 11µm thick, while the accessions held in Svalbard are held in 20µm laminated aluminium foils. Seed moisture content or eRH should be measured periodically to confirm that storage moisture is adequately maintained.

The storage temperature defines the maximum longevity possible for a seed sample and a stable storage environment is critical to maintaining seed viability. However, there are limited data from long-term storage at a range of low temperatures. Storage at −18 °C has been recommended in the past for long-term storage, as it is the lowest temperature that can be achieved with a single stage standard deep freezer compressor. For long-term stored seeds, all attempts should be made to maintain storage temperatures within ±3 °C of the set temperature and to limit the total duration of fluctuations outside this range to less than one week per year. Genebanks should maintain records of storage temperature deviations and periods when seed accessions are removed from the storage environment. For short-term storage, the seeds should be dried at the same temperature as they are stored, e.g. if ambient condition is 20 °C, seeds should then be dried at that same temperature.

Contingencies

Seeds in long-term storage should be removed rarely and only when samples in medium-term storage are exhausted or seeds need monitoring. Desired storage conditions are not achieved when mechanical environmental controls fail or when seeds are repeatedly removed from controlled storage environment. Backup generators with an adequate fuel supply should be available on-site.

All containers leak and seed moisture will eventually equilibrate to environmental conditions within the storage vault. This occurs faster in containers for which thermal plastics are used, as the moisture barrier is not absolute, or if glass or foil laminate containers have faulty seals or imperfections. Seeds may need to be re-dried occasionally and containers or gaskets replaced within 20–40 years.

If clear containers are used, perforated transparent plastic sachets containing self-indicating silica gel, equilibrated to the drying environment, can be used to monitor container performance during long-term storage. A change in colour of the silica gel inside the sachet (stored alongside the seeds) will indicate moisture ingress if the container seal fails. Orthodox seeds with short life spans or seeds with low initial quality may deteriorate more rapidly in storage and not meet long-term storage standards unless cryogenic conditions are used.

SELECTED REFERENCES

Dickie, J.B., Ellis, R.H., Kraak, H.L., Ryder, K. & Tompsett, P.B. 1990. Temperature and seed storage longevity. *Annals of Botany*, 65: 197–204.

Ellis, R.H. & Hong, T.D. 2006. Temperature sensitivity of the low-moisture-content limit to negative seed longevity-moisture content relationships in hermetic storage. *Annals of Botany*, 97: 785–91.

Ellis, R.H. & Roberts, E.H. 1980. Improved equations for the prediction of seed longevity. *Annals of Botany*, 45: 13–30.

Ellis, R.H., Hong, T.D. & Roberts, E.H. 1985. Sequential germination test plans and summary of preferred germination test procedures. *Handbook of seed technology for genebanks*. Vol I . Principles and methodology. Chapter 15, pp. 179–206. Rome, IBPGR.

Engels, J.M.M. & Visser, L., eds. 2003. *A guide to effective management of germplasm collections*. Handbooks for Genebanks No. 6. Rome, IPGRI.

Gómez-Campo, C. 2006. Erosion of genetic resources within seedbanks: the role of seed containers. *Seed Science Research*, 16: 291–294.

Harrington, J.F. 1972. Seed storage longevity. *In* T.T. Kozlowski, ed. *Seed biology*, Vol. III. pp. 145–245. New York, USA, Academic Press.

Nagel, M. & Börner, A. 2010. The longevity of crop seeds stored under ambient conditions. *Seed Science Research*, 20: 1–12.

Pérez-García, F., Gómez-Campo, C. & Ellis, R.H. 2009. Successful long-term ultra dry storage of seed of 15 species of Brassicaceae in a genebank: variation in ability to germinate over 40 years and dormancy. *Seed Science and Technology*, 37(3): 640–649.

Probert, R.J., Daws, M.I. & Hay, F.R. 2009. Ecological Correlates of Ex Situ Seed Longevity: a Comparative Study on 195 Species. *Annals of Botany*, 104 (1): 57–69.

Royal Botanic Gardens, Kew. Seed Information Database (SID). Predict seed viability module (available at: http://data.kew.org/sid/viability/percent1.jsp). Convert RH to water content (available at: http://data.kew.org/sid/viability/mc1.jsp). Convert water content to RH (available at: http://data.kew.org/sid/viability/rh.jsp). Kew, UK.

Smith, R.D., Dickie, J.B., Linington, S.H., Pritchard, H.W. & Probert, R.J., eds. 2003. *Seed conservation: turning science into practice*. Chapters 17 and 24. Kew, UK, Royal Botanic Gardens (available at: http://www.kew.org/msbp/scitech/publications/sctsip.htm).

Vertucci, C.W. & Roos, E.E. 1990. Theoretical basis of protocols for seed storage. *Plant Physiology*, 94: 1019–1023.

Walters, C. 1998. Understanding the mechanisms and kinetics of seed aging. *Seed Science Research*, 8: 223–244.

Walters, C. 2007. Materials used for seed storage containers. *Seed Science Research*, 17: 233–242.

Walters, C. & Engels, J. 1998. The effect of storing seeds under extremely dry conditions. *Seed Science Research*, 8. Supplement 1, pp. 3–8.

Walters, C., Wheeler, L.J. & Grotenhuis, J. 2005. Longevity of seeds stored in a genebank: species characteristics. *Seed Science Research*, 15: 1–20.

Walters, C., Wheeler, L.J. & Stanwood, P.C. 2004. Longevity of cryogenically-stored seeds. *Cryobiology*, 48: 229–244.

4.3 Standards for seed viability monitoring

Standards

4.3.1   The initial seed viability test should be conducted after cleaning and drying the accession or at the latest within 12 months after receipt of the sample at the genebank.

4.3.2   The initial germination value should exceed 85 percent for most seeds of cultivated crop species. For some specific accessions and wild and forest species that do not normally reach high levels of germination, a lower percentage could be accepted.

4.3.3   Viability monitoring test intervals should be set at one-third of the time predicted for viability to fall to 85 percent[1] of initial viability or lower depending on the species or specific accessions, but no longer than 40 years. If this deterioration period cannot be estimated and accessions are being held in long-term storage at – 18°C in hermetically closed containers, the interval should be ten years for species expected to be long-lived and five years or less for species expected to be short-lived.

4.3.4   The viability threshold for regeneration or other management decision such as recollection should be 85 percent or lower depending on the species or specific accessions of initial viability.

---

1   The time for seed viability to fall can be predicted for a range of crop species using an online application based on the Ellis/Roberts viability equations (see http://data.kew.org/sid/viability/).

Context

Good seed storage conditions maintain germplasm viability, but even under excellent conditions viability declines with period of storage. It is, therefore, necessary to assess viability periodically. The initial viability test should be conducted as early as possible before the seeds are packaged and enter the storage, and subsequent tests are conducted at intervals during storage. If for practical reasons of workflow and efficiency the initial viability test cannot be made prior to storage, it should be made as soon as possible and not later than 12 months after receiving. This can be the case of multi-species genebanks, where a wide range of germination regimes is required and samples of the same species are tested all together once a year.

The purpose of viability monitoring is to detect loss in viability during long-term storage before viability has fallen below the threshold for regeneration. The important guiding principle is one of active management of the collection. Too frequent monitoring will result in unnecessary waste of seeds and resources. On the other hand, significant viability decline may not be detected if monitoring is delayed or infrequent; advanced aging of the sample may result in genetic changes (random or directed selection), unrepaired mutations fixed in the sample, or ultimate loss of the accession.

When it is predicted that viability will fall to 85 percent before the next scheduled retest, the time of the retest should be anticipated or the accession directly scheduled for regeneration.

Risk of genetic erosion during storage is lower for homogeneous samples and germination. Decline to less than 85 percent is allowable as long as plant establishment during regeneration remains adequate. For heterogeneous samples such as wild species and landraces, the 85 percent standard should be adhered to. For some landraces, specific accessions, wild species and forest species, a viability of 85 percent in newly replenished seed is rarely achievable. In these situations, the curator can set the viability standard trigger for selected species to a lower threshold, such as 70 percent or lower.

Models to predict seed longevity from ambient to freezer conditions are available for diverse agricultural species. Genebank staff should use available predictive tools documented for particular species and storage conditions to anticipate duration that seeds will maintain high viability and to guide other genebank operations such as viability monitoring and regeneration frequencies (see Standards for viability monitoring and regeneration). Longevity predictions based on general species characteristics should be considered as estimates with large confidence intervals. Genebanks are encouraged to develop and report new information that describes and updates species responses to storage conditions.

Technical aspects

Viability monitoring intervals should be adjusted according to the data received from germination tests. As soon as a significant decline is detected, monitoring intervals should be reduced in order to 'fine tune' the prediction of time to reach the viability standard.

Accessions with very high initial viability (> 98 percent) may show a statistically significant decline in viability long before the predicted time for viability to fall to 85 percent, when germination is still well above 90 percent. Regeneration or recollection at this point is probably too soon and unnecessary. However, future retest intervals should be brought forward (e.g. from ten years to five years) in order to track the decline more accurately.

For accessions of lower quality, the accession might be dangerously close to the tipping point if viability declines comparatively rapidly. Such accessions should be managed carefully and the first viability monitoring tests should be after 3-5 years of storage intervals at first. Infrequent (e.g. ten-year) monitoring might fail to detect rapid deterioration and the viability threshold of 85 percent could be missed with negative consequences to the genetic integrity of the collection. In this respect, the use of statistical models can help to predict the tipping point and predict a time frame for appropriate regeneration.

Viability testing should give the manager an approximation of the viability of the sample. The goal should be to detect differences of +5 percent or so, rather than differences of +0.1 percent. Sample sizes for viability monitoring will inevitably be dependent upon the size of the accession but should be maximized to achieve statistical certainty. However, the sample size should be minimized to avoid wasting seed. Seed in a genebank is a valuable resource and should not be wasted.

It is difficult to establish a strict standard for the number of seeds for germination tests in genebanks. However, standard protocols as outlined by the International Seed Testing Association (ISTA) are often used. As a general guideline, 200 seeds are recommended to be used for initial germination tests (ISTA, 2008). If the initial germination is less than 90 percent, the sequential testing procedure proposed by Ellis *et al.* (1985) can help to save seed in subsequent germination tests during storage. However, in the event that there are not sufficient seeds, 100 or even smaller seed samples are also adequate and should be conducted with replications. The germination test is a guide of viability and even small seed samples can give the manager useful information. But in practice the actual sample size for germination will depend on the size of the accession, which is very limited in general (ideally the recommended minimum size for self-pollinated is 1 500 and for cross-pollinated species 3 000 seeds) in genebanks. It is important to minimize the use of valuable seeds required for germination tests. For small accession sizes (as is often the case for wild species), sample sizes of 50 seeds or less could be acceptable. However, it must be realized then that there may be a higher chance of germination being below the threshold. The genebank curator should assess the risk of this occurrence.

The germination test should always be used in preference to alternatives such as the tetrazolium test. However, in circumstances where it is not possible to remove seed dormancy, alternative tests may be carried out. It is recommended that germination be measured at two different times to have an idea of fast and slow germinating seeds. Records of the number of abnormally germinating seeds should also be kept. Slower germination and increasing abnormals are often early indicators that deterioration is occurring.

Every effort should be made to germinate all viable seeds in a collection using optimum conditions and appropriate dormancy-breaking treatments where needed. Non-germinated seeds remaining at the end of a germination test should be cut-tested to assess whether they are dead or dormant. Seeds with firm, fresh tissue are likely to be dormant and should be counted as viable seeds.

All data and information generated during viability monitoring should be recorded and entered into the documentation system.

Contingencies

It is recognized that viability monitoring is an expensive activity and that genebanks would wish to seek cost-cutting procedures. One such procedure may entail measuring seed quality in a subsample of accessions of the same species grown in the same harvest year. This practice may reveal overall trends on the effect of harvest year on seed quality, but will not take genotype by harvest year interactions into consideration that are known to be important for seed quality.

Where different harvest conditions occur over a wide range of maturities across accessions, then a sampling strategy can be from separate harvested sub-groups. An additional strategy would be to focus retesting on the accessions that gave the lowest viability result in the initial tests. Retest data from these accessions should provide an early warning on the performance of the batch as a whole.

The initial germination test at harvest for known hard-seeded species and accessions frequently found in some forage legume species and Crop Wild Relatives can be as low as 45 percent, and increases after 10–15 years to 95 percent or more and remains so for long periods. If the initial germination is less than 90 percent, then regenerate/recollect at first detectable significant decline established by an appropriate statistical test.

However, it is recognized that intraspecific variation among accessions has been observed for a wide range of accessions, thus, there are risks associated with the above strategies, which should be considered. Viability monitoring of accessions of wild species is generally more problematic compared with crop species. Seed dormancy is likely to be much more prevalent and accession sizes are often small meaning that smaller minimum sample sizes have to be adopted for germination tests, as this will inevitably affect the ability to detect the onset of seed deterioration.

With reference to the initial seed viability testing, it is also possible that genebanks receive small quantities of seeds. In that case, it is not necessary to carry out initial seed viability testing since the sample is sent for the purpose of regeneration. However, the regenerated seeds must then be tested for viability prior to storage.

The range of inherent longevity is also wider in wild species with some species from Mediterranean and tropical dryland habitats expected to be extremely long-lived and, conversely, some species from cold, temperate regions expected to be short-lived. For the latter, retesting intervals of as few as three years should be considered as well as duplication into cryostorage as a precautionary measure. In the event that storage conditions are not met (as will occur if there is a prolonged power cut when seeds are stored in refrigeration units), viability will be affected negatively depending on the species, length of disruption and conditions during the disruption. In such an event, a disaster management plan should be activated. For example, some representative samples may need to be tested immediately following resumption of adequate storage conditions.

SELECTED REFERENCES

AOSA (Association of Official Seed Analysts). 2005. Page 113 in Capashew, ed. *Rules for testing seeds*, 4-0, 4-11. Las Cruces, New Mexico, USA.

Dickie, J.B., Ellis, R.H., Kraak, H.L., Ryder, K. & Tompsett, P.B. 1990. Temperature and seed storage longevity. *Annals of Botany*, 65: 197–204.

Ellis, R.H. & Roberts, E.H. 1980. Improved equations for the prediction of seed longevity. *Annals of Botany*, 45: 13–30.

Ellis, R.H., Hong, T.D. & Roberts, E.H. 1985. Sequential germination test plans and summary of preferred germination test procedures. *Handbook of seed technology for genebanks. Vol I . Principles and methodology.* Chapter 15, pp 179–206. Rome, IBPGR.

Engels, J.M.M. & Visser, L., eds. 2003. *A guide to effective management of germplasm collections.* Handbooks for Genebanks No. 6. Rome, IPGRI.

ENSCONET. 2009. Manuals (available at: http://ensconet.maich.gr/Download.htm).

Harrington, J.F. 1972. Seed storage longevity. *In* T.T. Kozlowski, ed. *Seed biology*, Vol III, pp. 145–245, New York, USA, Academic Press.

ISTA (International Seed Testing Association). 2008. *International rules for seed testing.* Bassersdorf, Switzerland.

Nagel, M. & Börner, A. 2010. The longevity of crop seeds stored under ambient conditions. *Seed Science Research*, 20: 1–12.

Nagel, M., Rehman Arif, M.A., Rosenhauer, M. & Börner, A. 2010. *Longevity of seeds - intraspecific differences in the Gatersleben genebank collections.* Tagungsband der 60. Jahrestagung der Vereinigung der Pflanzenzüchter und Saatgutkaufleute Österreichs 2009, 179–181.

Royal Botanical Gardens, Kew. Seed information database (SID). Kew, UK (available at: http://data.kew.org/sid/).

Smith, R.D., Dickie, J.B., Linington, S.H., Pritchard, H.W. & Probert, R.J., eds. 2003. *Seed conservation: turning science into practice.* Chapters 17 and 24. Kew, UK, Royal Botanic Gardens (available at: http://www.kew.org/msbp/scitech/publications/sctsip.htm).

4.4 Standards for regeneration

Standards

4.4.1 Regeneration should be carried when the viability drops below 85 percent of the initial viability or when the remaining seed quantity is less than what is required for three sowings of a representative population of the accession. The most-original-sample should be used to regenerate those accessions.

4.4.2 The regeneration should be carried out in such a manner that the genetic integrity of a given accession is maintained. Species-specific regeneration measures should be taken to prevent admixtures or genetic contamination arising from pollen geneflow that originated from other accessions of the same species or from other species around the regeneration fields.

4.4.3 If possible at least 50 seeds of the original and the subsequent most-original-samples should be archived in long-term storage for reference purposes.

Context

Regeneration is a key operation and an integral responsibility of any genebank that maintains orthodox seeds. It is a process that leads to an increase of the stored seeds (also called "multiplication") in the genebank and/or to an increase of the viability of the seeds equal to or above an agreed minimum level, which is referred to as the regeneration threshold. An accession will be regenerated when it does not have sufficient seeds for long-term storage (e.g. 1 500 seeds for a self-pollinating species and 3 000 for an out-crossing species) or when its viability has dropped below an established minimum threshold (i.e. below 85 percent of initial viability of the stored seeds). Regeneration should also occur when the seed numbers have been depleted due to frequent use of the accession. If an accession is rarely requested and seed viability is fine, then seed numbers can be below 1 000 prior to regeneration, each regeneration of especially out-crossing species runs the risk of losing rare alleles or changing the genetic profile for the sample. Regeneration frequency should be minimized. High seed numbers are not needed for rarely requested accessions or species.

As regeneration is an activity that could easily affect the genetic composition of an accession (and thus its genetic integrity) utmost care is required. Consequently, genebank operators will have to strike a delicate balance between avoiding regeneration as much as possible versus the potential loss of viability and thus, the risk of affecting the genetic integrity of an accession. Active management of the collections will greatly help to decide on the best moment to regenerate.

Regeneration should be undertaken with the least possible change to the genetic integrity of the accession in question. This means that, in addition to sampling considerations (see paragraph below) of the accession in question, due attention should be paid to the environment in which the activity will be undertaken, to avoid any severe selection pressure on the accession. It has been suggested that the regeneration environment should be as similar as possible to that at the collecting site, in particular when a population collected in the wild is being regenerated, in order to minimize genetic drift and shift as well as to produce the best possible quality of seeds. It can often be difficult to harvest sufficient quantity of seed from wild relatives due to lower seed/plant numbers compared to other species, or plant dispersal mechanisms such as seed shattering. It is therefore necessary to ensure that appropriate technical practices are used to capture as much seed as possible (i.e. nets to capture dropped seeds).

Repeat regeneration cycles may also be required to ensure that sufficient seed is conserved. For regeneration, it is better to create favourable environmental conditions for seed production and minimize plant-to-plant competition. Conditions at the original collection sites are often unfavourable in one or more ways for maximizing seed production. So there should really be a compromise between generalized, favourable conditions and those special signals (whether photoperiodic, nutritional or climatic) that are specific to local adaptation of individual accessions. This is part of the art of curation. If the genebank site does not provide favourable conditions locally, a curator should explore means to have the collection regenerated in a favourable environment; replication of the collection environment should not necessarily be the curator's goal.

To preserve the genetic integrity of genebank collections during seed regeneration, it is important that sampling of accessions be carried out effectively. The number of seeds to be used for the regeneration process must be sufficient to be representative of the genetic diversity in an accession and to capture one or more rare alleles with a certain probability.

The methodology to be used for regeneration might vary from species to species and depends, among other factors, on the population size, breeding system and pollination efficacy. Therefore, it is of significant importance to collate as much as possible of the relevant biological information related to the species in question. In addition, when possible and meaningful, it is recommended that the regeneration event be used also for the characterization of regenerated accessions (see Characterization Standards). However for cross pollinating species, it is often difficult, to use the regeneration process to carry out characterization due to logistical reasons.

Technical aspects

In order to maintain the genetic integrity of accessions it is recommended to use seeds from the most-original-sample for regeneration. For multiplication, it is recommended to use seeds from the working collection for up to five cycles of multiplication without returning to the most-original-sample.

It should be noted that in cases where the original collection or donation is a small sample, it is necessary to regenerate immediately following receipt of the material in order to obtain an adequate quantity of seeds for long-term storage. It is important to record the number of the regeneration cycle and enter the information into the documentation system. It is recommended that the receiving genebank always keep some seeds from the initial seed sample for future reference purposes. Even if these original seeds lose their viability, they can be useful in confirming morphology or genotype of later generations of the respective accession.

The size of the seed sample to be used in the regeneration activity has to reflect the genetic composition of the accession. For this purpose, the effective population size (N_e) is a key parameter that will have a bearing on the degree of genetic drift that is associated with the regeneration of the accession. This minimal size of N_e to minimize loss of alleles can be estimated for individual accessions based on the pollination biology and growing conditions. Best practices for harvesting should be used to avoid seed mixture during seeding, harvest and processing. Research by Johnson *et al.* (2002, 2004) on the regeneration of perennial allogamous species (e.g.

grasses) indicated that 100 plants is a minimum number which is necessary for the preservation of taxon gene pool. The principle of harvesting from 3 to 5 inflorescences from each plant is recommended.

To avoid geneflow/contamination it is critically important to use proper isolation methods between plots of accessions of cross-pollinated species being regenerated. This also applies to self-pollinated species, depending on the regeneration environment. For species that depend on specific pollinators, isolation cages and the corresponding pollinators should be used (Dulloo *et al.*, 2008). Contamination and genetic drift/shift can be assessed with morphological, enzymatic or other distinctive traits that can be used as markers (e.g. flower colour; seed colour), or with molecular markers.

Reference collections (herbarium specimen, photographs and/or descriptions of the original accessions) are essential for conducting the true-to-type verification (Lehmann and Mansfeld, 1957). Close inspections of obtained seeds and during the first regeneration of a new genebank accession are required to collect important reference information. In order to avoid differences in seed maturity in a seed sample, multiple harvests should be carried out during the fruiting season.

Contingencies

There will always be a risk management dimension to the curatorship role. Solid biological knowledge of the species in question is a key factor in making the best possible decisions for regeneration under constrained conditions. Aspects such as sample size, distance between individual accessions and other forms of isolating accessions, respecting established thresholds for viability loss, growing conditions and others, all need to be given due attention when planning the regeneration activity.

In view of this complexity, it is not meaningful to look for possible contingencies. In case of emergency, it would be advisable to seek advice from experts and/or collaboration with other genebanks that could provide assistance.

SELECTED REFERENCES

Breese, E.L. 1989. *Regeneration and multiplication of germplasm resources in seed genebanks: the scientific background*. Rome, IBPGR.

Crossa, J. 1995. Sample size and effective population size in seed regeneration of monecious species. *In* J.M.M. Engels & R. Rao, eds. *Regeneration of seed crops and their wild relatives*, pp. 140–143. Proceedings of a consultation meeting, 4-7 December 1995. Hyderabad, India, ICRISAT, and Rome, IPGRI.

Dulloo, M.E., Hanson, J., Jorge, M.A. & Thormann, I. 2008. Regeneration guidelines: general guiding principles. *In* M.E. Dulloo, I. Thormann, M.A. Jorge & J. Hanson, eds. *Crop specific regeneration guidelines*. [CD-Rom], Rome, SGRP-CGIAR.

Engels, J.M.M. & Rao, R., eds. 1995. *Regeneration of seed crops and their wild relatives*, pp. 140–143. Proceedings of a consultation meeting, 4-7 December 1995. Hyderabad, India, ICRISAT, and Rome, IPGRI.

Engels, J.M.M. & Visser, L., eds. 2003. *A guide to effective management of germplasm collections*. Handbooks for Genebanks No. 6. Rome, IPGRI.

Johnson, R.C., Bradley, V.L., & Evans, M.A. 2002. Effective population size during grass germplasm seed regeneration. *Crop Science*, 42: 286–290.

Johnson, R.C., Bradley, V.L., & Evans, M.A. 2004. Inflorescence sampling improves effective population size of grasses. *Crop Science*, 44: 1450–1455.

Lawrence, L. 2002. A comprehensive collection and regeneration strategy for *ex situ* conservation. *Genetic resources and crop evolution*, 49(2): 199–209.

Lehmann, C.O. & Mansfeld, R. 1957. Zur Technik der Sortimentserhaltung. *Kulturpflanze*, 5: 108–138.

Rao, N.K., Hanson, J., Dulloo, M.E., Ghosh, K., Nowell, D. & Larinde, M. 2006. *Manual of seed handling in genebanks*. Handbooks for Genebanks No. 8. Rome, Bioversity International.

Sackville Hamilton, N.R. & Chorlton, K.H. 1997. *Regeneration of accessions in seed collections: a decision guide*. J. Engels, ed. Handbooks for Genebanks No. 5. Rome, IPGRI.

SGRP-CGIAR. Crop Genebank Knowledge Base (available at: http://cropgenebank.sgrp.cgiar.org).

4.9   Standards for safety duplication

Standards

4.9.1   A safety duplicate sample for every original accession should be stored in a geographically distant area, under the same or better conditions than those in the original genebank.

4.9.2   Each safety duplicate sample should be accompanied by relevant associated information.

Context

Safety duplication assures the availability of a genetically identical subsample of the accession to mitigate the risk of its partial or total loss caused by natural or human-caused catastrophes. The safety duplicates are genetically identical to the long-term collection and are referred to as the secondary most-original-sample (Engels and Visser, 2003). Safety duplication includes both the duplication of material and its related information, including database backup. The safety duplication of the materials is deposited in long-term storage at a different location. The location is chosen to minimize possible risks and provides the best possible storage facilities. To minimize risks that can arise in any individual country safety duplication will be ideally undertaken outside that country.

Safety duplication is generally made under a 'black-box' approach. This means that the repository genebank has no entitlement to the use and distribution of the germplasm. It is the depositor's responsibility to ensure that the deposited material is of high quality, to monitor seed viability over time and to use their own base collection to regenerate the collections when they begin to lose viability. The germplasm is not touched without permission from the depositor and is only returned on request when the original collection is lost or destroyed. Recall of the deposit is also possible when it is replaced with newly regenerated germplasm. It is recognized however that the black-box is not the only approach. There may be cases where the safety collection is also taken care of by the recipient genebank.

Safety duplication should be made for all original seeds collected by the genebank or when only held by the genebank. However, the genebank should retain a set of the original samples to facilitate access for regeneration or other managerial decisions. Seeds that are duplicates from other collections can usually be retrieved from those collections and do not require safety duplication unless there is doubt about their security in the other collection.

Any safety duplication arrangement requires a clearly signed legal agreement between the depositor and the recipient of the safety duplicate that sets out the responsibilities of the parties and terms and conditions under which the material is maintained.

Safety duplication is available at the Svalbard Global Seed Vault on Spitsbergen Island, Norway. Institutions depositing seeds retain ownership and access to samples stored in Svalbard is granted to the depositor only.

Technical aspects

When selecting the location for safety duplication, primary consideration is given to the geographic location and environmental conditions of the location. Facilities must ensure low radiation (radioactivity) and stability (low probability of earthquakes). The facility must be situated at an elevation that guarantees proper drainage during seasonal rains and eliminates the risk of flooding in the event of rising sea levels due to global warming. Equally important is economic stability and socio-political certainty. Koo *et al.* (2004) suggest that safety duplicate samples should be located away from the risk of political embargo, military action or terrorism that could disrupt international access.

Samples are prepared for safety duplication in the same way as for the base collection. Conditions should be at least as stringent as those for long-term storage of germplasm in a genebank and the quality of seed preparation (i.e. drying) is important. In some cases, it is helpful to sort material according to short, medium and long living seed groups before sending for safety duplication.

Sample size should not be restricted to a certain minimum number. Sample size should be sufficient to conduct at least three regenerations. A safety backup is not just for future regeneration; it may also provide a minimum sample to regenerate an accession that was lost. A "critical" safety backup with a minimal amount of seed at a second location is better than no backup at all. If possible, a safety duplicate of an accession in a seed genebank should contain at least 500 viable seeds for outbreeders and heterogeneous accessions with high diversity and a minimum of 300 seeds for genetically uniform accessions. For accessions with seeds of low viability more seeds are necessary. Storage temperatures should be –18 °C to –20 °C.

The packaging material for safety duplication should be of trilaminate material of which the middle metal foil layer should be of adequate thickness. It should be formed into a pouch seamed on all four sides with no gusset. This would provide an adequate water barrier for transport and storage at 18 °C for at least 30 years. An outer and inner label should be placed on each packet of seeds to ensure that the germplasm is properly identified.

As the storage conditions for the safety duplicate should be the same or better than that of the base collection, seed viability can be monitored on seed lots of the same accession maintained in long-term storage in the genebank and extrapolated to the safety duplicate if basic standards for storage conditions are met and the same containers are used. In some cases, samples for germination testing may be sent in a separate box with the safety duplicate and monitored for germination by agreement with the depository.

Strong cold-resistant boxes (cardboards or polypropylene boxes) are the best options for transporting and storing seeds. Boxes should be sealed properly. Shipment should consider the fastest means of transport available either by air freight, courier or by land to avoid deterioration of seed quality during transit. Samples should be renewed from the sender when the viability of the samples in similar storage conditions in the long-term collection of the sender starts to decline.

Contingencies

When extrapolating the viability of the safety duplicate from viability monitoring results of the sample in the base collection, some caution should however be taken. Seeds may age at different rates if there is a difference in ambient RH at the two sites and/or differences in extent or frequency of temperature fluctuations, though the average storage temperature is the same.

Issues of liability may occur related to sending samples in sealed black-box conditions. One issue is on liability for contents of the sealed box and handling by customs officers and other authorities for entry into a country. In some cases, boxes are opened and special seals are applied by the authorities to confirm that the samples are not medicinal or other prohibited plants. Another issue is that on liability of the recipient institution should material be damaged or lose viability earlier than expected as a result of stress during transit, faulty seal of containers, or temperatures that fluctuate from specified standards. Under the conditions described here, the safety duplicate repository should only be "liable" if the temperature becomes uncontrollable; this should be reported immediately to the primary institution so that they can decide on what action to take. The primary institution should bear full responsibility for transport disasters or uncontrolled moisture.

The standards and technical aspects may be difficult to implement for some species due to the inherent biology of the samples, e.g. short-lived seeds, large-seeded species where space and cost may be limiting.

SELECTED REFERENCES

Engels, J.M.M. & Visser, L., eds. 2003. *A guide to effective management of germplasm collections.* Handbooks for Genebanks No. 6. Rome, IPGRI.

Koo, B., Pardey, P.G., Wright, B.D., Bramel, P., Debouck, D., Van Dusen, M.E., Jackson, M.T., Rao, N.K., Skovmand, B., Taba, S. & Valkoun, J. 2004. *Saving seeds: The economics of conserving crop genetic resources ex situ in the future harvest centres of the CGIAR.* Wallingford, UK, CAB International.

SGRP-CGIAR. Crop Genebank Knowledge Base. Page on safety duplication: Background documents, list of references and standard safety deposit agreement template (available at: http://cropgenebank. sgrp.cgiar.org/index.php?option=com_content&view=article&id=58&Itemid=207&lang=English).

4.10 Standards for security and personnel

Standards

4.10.1 A genebank should have a risk management strategy in place that includes *inter alia* measures against power cut, fire, flooding and earthquakes.

4.10.2 A genebank should follow the local Occupational Safety and Health requirements and protocols where applicable.

4.10.3 A genebank should employ the requisite staff to fulfil all the routine responsibilities to ensure that the genebank can acquire, conserve and distribute germplasm according to the standards.

Context

Achieving a genebank's goal of acquisition, conservation and distribution of germplasm not only require adequate procedures and equipment for germplasm handling be in place, but that properly trained staff be employed to carry out the required work and to guarantee the security of the genebank.

Active genebank management requires well-trained staff, and it is crucial to allocate responsibilities to suitably competent employees. A genebank should, therefore, have a plan or strategy in place for personnel, and a corresponding budget so as to guarantee that a minimum of properly trained personnel is available to fulfil the responsibilities of ensuring that the genebank can acquire, conserve and distribute germplasm. Access to specialists in a range of subject areas is desirable, depending on the mandate and objectives of each individual genebank. However, staff complements and training will depend on specific circumstances. The health and usefulness of the seeds stored in the genebank depend also on issues related to safety and security of the genebank. Arrangements need to be in place for electricity backup; fire extinction equipment has to be in place and regularly checked; genebank buildings need to be earthquake-proof if situated in a seismic-prone area, to mention some. A genebank should, therefore, implement and promote systematic risk management that addresses the physical and biological risks in the every-day environment to which the collections and related information are exposed.

Technical aspects

Staff should have adequate training acquired through certified training and/or on-the-job training and training needs should be analyzed. Genebank personnel should be aware of and trained in safety procedures to minimize risks to the germplasm.

The genebank facilities should be constructed so as to withstand natural disasters, such as hurricanes, cyclones, earthquakes, or floods that are known to occur in the location where the genebank has been built.

Storage facilities should be protected with standard security facilities such as fences, alarm systems, security doors and any other system that helps to shield the genebank from burglars and other intruders. Security of the seed collections in the genebank will be enhanced by allowing entry strictly to authorized personnel into the actual storage facilities.

Protective clothing should be provided and used in the storage area. Adequate precautions should be taken and safety equipment, including alarms and devices to open doors from inside drying rooms and refrigerated rooms, should be installed.

Refrigeration will almost certainly be reliant on electrical power and it is, therefore, necessary that the power supply is adequate and reliable. Failure in power supply can result in complete loss of genebank accessions. Consideration should be given to the provision of a backup generator that automatically cuts in when the main power supply fails. This will require stockpiling adequate amounts of fuel to run the generator during power cuts.

Monitoring devices for temperature should be available in the drying and storage rooms to track the actual parameters against time. It should be considered whether it is better to store seed without refrigeration if refrigeration is inherently unreliable. If refrigeration is to be used to conserve germplasm, it must meet necessary standards, as unreliable refrigeration can be far more damaging than non-refrigerated storage.

If refrigeration and/or electric power are unreliable, a facility can be built in the soil at a depth of 10–20 m, where temperature can be averaged at 10 °C. This could be attractive in several tropical regions under no risk of flooding. Drying should be well carried out however, and seeds should be kept in properly sealed vials.

Fire alarm and fire-fighting equipment is required in the genebank. Most fires begin from faulty electrical circuits and periodic checks should be made on the electrical circuitry to ensure compliance with safety standards. Fire-fighting equipment will include extinguishers and fire blankets. For areas affected by thunderstorms, a lightning rod should be fitted to the genebank.

Contingencies

When suitably trained staff is not available, or when there are time or other constraints, it might be a solution to outsource some of the genebank work or to approach other genebanks for assistance. The international community of genebanks should be informed, if the functions of the genebank are endangered.

Unauthorized entry to genebank facilities can result in direct loss of material, but can also jeopardize the collections through inadvertent introduction of pests and diseases and interference in management systems.

SELECTED REFERENCES

Engels, J.M.M. & Visser, L., eds. 2003. *A guide to effective management of germplasm collections.* Handbooks for Genebanks No. 6. Rome, IPGRI.

SGRP-CGIAR. Crop Genebank Knowledge Base. Page on risk management (available at: http://cropgenebank.sgrp.cgiar.org/index.php?option=com_content&view=article&id=135&Itemid=236&lang=english).

Chapter 5

Field genebank standards

5.1 Standards for choice of location of the field genebank

Standards

5.1.1 The agro-ecological conditions (climate, elevation, soil, drainage) of the field genebank site should be as similar as possible to the environment where the collected plant materials were normally grown or collected.

5.1.2 The site of the field genebank should be located so as to minimize risks from natural and manmade disasters and hazards such as pests, diseases, animal damage, floods, droughts, fires, snow and freeze damage, volcanoes, hails, thefts or vandals.

5.1.3 For those species that are used to produce seeds for distribution, the site of the field genebank should be located so as, to minimize risks of geneflow and contamination from crops or wild populations of the same species to maintain genetic integrity.

5.1.4 The site of the field genebank should have a secured land tenure and should be large enough to allow for future expansion of the collection.

5.1.5 The site of the field genebank should be easily accessible to staff and supplies deliveries and have easy access to water, and adequate facilities for propagation and quarantine.

Context

Considering the long-term nature of a field genebank, the selection of an appropriate site for its location is critical for the successful conservation of germplasm. There are many factors that need to be taken into account when selecting a site for a field genebank including appropriate agro-ecological condition for the plants being conserved at the site, associated natural and manmade disasters, secure long-term land tenure, accessibility of the site for staff and availability of water resources.

Technical aspects

Plants will grow strong and healthy when planted under appropriate agro-ecological conditions. Field genebanks are particularly vulnerable to losses caused by poor adaption of material that has originated in environments that are very different from that of the genebank location. The selected site for the field genebank should have an environment and soil type best suited for the species to reduce the risk of poor adaptation. One solution to poor adaptation is to take a decentralized approach to genebank management, i.e. to collocate the collections in different agro-ecologies rather than in a centralized genebank. Accessions of similar adaptation are kept together in a station located in an agro-environment similar to their origin or similar or near to their natural habitat. The natural conditions of the original environment can be simulated by providing higher shade intensity or drainage, for example for crop wild relatives that originated in natural forests versus cultivated plants that are adapted to higher light intensity.

Avoidance of pests and diseases and insect vectors are very important for field collections. If possible, the field genebank should be located in a location that is free from major pathogenic diseases and pests or away from known infected regions for fungi and virus to reduce risk and management costs related to plant protection and ensure a clean source of material for distribution. Soils should be checked before planting to ensure they are free from fungi, termites or other soil-borne parasites and appropriate treatment provided to clean soil before planting. Where this is not possible, the selected site should be located at some distance from fields of the same crop to reduce threats from insect pests and diseases and diseased plants should be removed with a vigorous roguing programme. If possible, maintain collections in areas with a hot and dry climate, which is less favourable for vector movement, pests and diseases. Further, the bringing together of large numbers of plants susceptible to disease may severely enhance the risk of disease outbreaks. Such large collections of single genera deserve particular scrutiny from a disease point of view.

The assessment of risk from natural disasters such as floods, fires, snow/ice, volcanoes, earthquakes and hurricanes is an important criterion for ensuring the physical safety of collections. In addition, physical security and potential of anthropogenic threats such as theft and vandalism should be taken into account. These characteristics should be considered when locating and designing a field genebank to help reduce loss of germplasm (see also Standards on safety).

Insect netting and cages can be used for protection against insect or bird damage for smaller plants. Out-crossing species such as fruit trees with recalcitrant seeds or grasses that are grown for seed as well as maintained as plants require isolation from potential pollinators. Selecting a site away from crop stands or wild populations of the same species to avoid gene flow or weed contamination is important for ensuring genetic integrity in these species. Recommended isolation distances, isolation cages or pollination control measures should be established and followed for propagation. Crop-specific information about isolation distance in regenerating accessions is available on the Crop Genebank Knowledge Base (see references).

A field genebank should be located in a secure site with a long-term agreement and guaranteed or gazetted land tenure and funding, taking into consideration the development plan for the area. The land-use history can give information about the pest or weed status of the land and the quantity of fertilizer used. High use of fertilizer in previous years could affect the growths of root and tubers. High residual fertilizer for example, can prevent tuber development in sweet potatoes. Drought stress can be avoided when the availability of adequate rainfall or water supply for supplementary irrigation is included as a selection criterion. Apart from land-use history, it is recommended to include measures that can be taken to ascertain and correct the physical and nutritional status of soils. This basically entails soil physical and chemical analysis followed by subsequent corrective measures. Areas with high potassium usage need to be balanced with supplemental calcium and magnesium applications, especially for tropical fruit trees.

The size of the chosen site should provide sufficient space for the type of species to be conserved as well as for possible future expansion when the collection grows, especially in the case of perennial species. Required space for tree crops can be considerable. Also, sufficient space should be available to accommodate annuals that require continuous replanting and rotation between plots to avoid any possible contamination from previous plantings, as well as rotation of annuals and perennials to control disease and manage soil fertility. Sufficient and appropriate storage facilities are required if plant material needs to be stored after harvest before the next planting.

Easy physical access to germplasm will aid monitoring and plant management. The site should be suitable for access of labour and machinery for mulching, fertilizer and pesticide applications and have access to adequate year-round irrigation, propagation, and *in vitro* or cryopreservation facilities as required. A good security system should be in place to avoid theft or damage to germplasm and facilities.

Contingencies

When accessions from different eco-geographical origins are planted in one location, careful attention by the curatorial field staff is required to monitor the reproductive phenology and seed production, and identify and transfer poorly adapted accessions to possible alternative sites, greenhouses, or *in vitro* culture to avoid genetic loss. Special management practices may be required for some accessions. Protected areas such as screenhouses or cages may be required to protect the plants from predators.

SELECTED REFERENCES

Anderson, C.M. 2000. *Citrus germplasm resources and their use in Argentina, Brazil, Chile, Cuba and Uruguay.* Proc. IX ISC. Vol I: 123–125, Orlando, Florida, USA.

Anderson, C.M. 2008. *Recursos genéticos y propagación de variedades comerciales de cítricos.* XII Simposium Internacional de Citricultura. Tamaulipas, México.

Borokini, T.I, Okere, A.U., Giwa, A.O., Daramola, B.O. & Odofin, T.W. 2010. Biodiversity and conservation of plant genetic resources in field genebank of National Centre for Genetic Resources and Biotechnology, Ibadan, Nigeria. *International Journal of Biodiversity and Conservation,* 2(3): 037–050.

Davies, F.S. & Albrigo, L.G. 1994. *Citrus.* Wallingford, UK, CAB International.

Gmitter, F.G. & Hu, X.L. 1990. The possible role of Yunnan, China, in the origin of contemporary citrus species (Rutaceae). *Economic Botany,* 44: 267–277.

Said Saad, M. & Rao, V.R., eds. 2001. *Establishment and management of field genebank training manual.* Serdang, Malaysia, IPGRI-APO.

SGRP-CGIAR. Crop Genebank Knowledge Base (available at: http://cropgenebank.sgrp.cgiar.org/).

5.2 Standards for acquisition of germplasm

Standards

5.2.1 All germplasm accessions added to the genebank should be legally acquired, with relevant technical documentation.

5.2.2 All material should be accompanied by at least a minimum of associated data as detailed in the FAO/Bioversity multi-crop passport descriptors.

5.2.3 Propagating material should be collected from healthy growing plants whenever possible, and at an adequate maturity stage to be suitable for propagation.

5.2.4 The period between collecting, shipping and processing and then transferring to the field genebank should be as short as possible to prevent loss and deterioration of the material.

5.2.5 Samples acquired from other countries or regions within the country should pass through the relevant quarantine process and meet the associated requirements before being incorporated into the field collection.

Context

Acquisition is the process of collecting or requesting such materials for inclusion in the field genebank, together with related information. The nature of plants with recalcitrant seed and vegetatively propagated plants requires special attention when acquiring germplasm for conservation in field genebanks. The propagules required for establishing a field genebank may come in different forms such as seeds, cuttings, tubers, corms, scionwood, tissue cultures, graftwood, or cryopreserved material. The plant materials may be obtained from existing genebanks, research and breeders' collections, landraces and cultivated forms grown by farmers and from plant explorations/expeditions. The relevant national and international regulations, such as phytosanitary/quarantine laws and national laws for genetic resources access, the IPPC, ITPGRFA, CBD, and any others that govern the movement and acquisition of germplasm, must be taken in to account.

Technical aspects

Adherence to Standard 5.2.1 will allow the safe movement of germplasm both from collection sites within the country and outside the country to the site hosting the genebank. When germplasm material is collected *in situ,* it is important to adhere to the national regulations, which normally require that collecting permits are obtained from relevant national authorities. If the collection is from farmers' fields or community areas prior informed consent may be required in accordance with relevant national, regional or international law. If germplasm material has to be exported from a country, an appropriate material transfer agreement should be used. In the case of PGRFA, the export can be accompanied with the SMTA or other similar permits in compliance with national regulations of access and benefit-sharing. Import permit regulations, which specify phytosanitary and any other import requirements, must be sought from the relevant national authority of the receiving country.

During the acquisition phase, it is important to ensure that passport data for each accession are as complete as possible. Especially, georeferenced data are very useful as they give a precise account of the location of the original collecting sites and help to identify accessions with specific adaptive traits in accordance to the agro-climatic conditions of the original collecting sites. Passport data are crucial in identifying and classifying each accession and will function as an entry point in selecting and using the accession. Appropriate collecting forms should be used to capture comprehensive collecting data. These forms should include information such as the initial taxonomic classification of the sample, the latitude and longitude of the collecting site, a description of the habitat of the collected plants, the number of plants sampled and other relevant data that are important for proper conservation, as provided in the FAO/Bioversity multi-crop passport descriptors (Alercia *et al.,* 2012). Very useful additional information, such as cultural practices, methods of propagation, history and origin, and uses can be obtained with interviews when material is collected from farmer fields. Whenever possible, a herbarium voucher

specimen collected from the same population as the samples, should be kept as a reference collection, and a record should be made of the method and reason for acquisition.

In the case of donations (from research programme or genebank), the taxonomic classification, donor name, donor identification number, and names of germplasm in addition to the available passport data should be provided. Adequate information about how the germplasm received was maintained, including pedigree or lineage information, as well as chain of custody information where available should be sought from the donor. Materials should be assigned a unique identification number (either temporary or permanent, according to the practice used in the genebank) that will link the material to the passport data and any other collected information, guaranteeing the authenticity of the sample.

Although it is not possible to ensure that plant material collected *in situ* is in completely healthy condition (no diseases and insect pest infestation) it is important that as far as possible propagules are collected from plants that appear healthy, devoid of disease and insect pest infestations or damage. Clean material acquired from certified sources should be stored in a screenhouse, to prevent insects from infesting clean plants and spreading pathogens. During collecting, the collector should also avoid the depletion of the natural population targeted for collecting. It may also be useful to repeat sampling from a particular site to maximize capture of genetic variability that may be present at various points in time (Guarino *et al.*, 1995). In the collection phase of vegetatively propagated perennial samples, especially when collecting shoots suitable for taking cuttings or grafting, it would be desirable to stimulate the formation of adequate shoots by scoring the trunk or the branches; these shoots could then be collected during a second visit.

It is important to highlight that the time taken to transfer the original genetic resource from the time of collecting to the genebank is critical. This is especially true for species that produce recalcitrant seeds and clonal stock, which do not retain their viability for very long and for vegetative propagules that decay easily. In some cases, germplasm material may need to be shipped over long distances, as the case may be when the material is acquired from other countries. Due consideration of the shipping period including transit and processing period, should be taken into account and appropriate measures taken to ensure that the material reaches the destination genebank in good condition. It is also important to properly prepare the propagules (scion woods, seeds or cuttings) to improve viability during postal or parcel transportation. For example, recalcitrant seeds and scions should be packed in sterile cotton or other suitable material in a perforated plastic bag to ensure sufficient air exchange. Seeds should be protected from crushing by mechanical mail sorter in rigid cushioned shipping. For scion wood, the two cut ends of the cleaned scion should be wrapped using a para-film strip to reduce moisture loss. Collections sent from tropical areas need to be mindful of high temperatures during transportation.

Given that field collections cannot accommodate many samples (see Standards for establishment of collection), the sample size for collecting will usually be limited compared to orthodox seeds. Nevertheless, all attempts should be made at maximizing the collection for the target population's genetic diversity. In addition, in collecting for a field genebank, the collector will need to take decisions on how many plants within a population can practically be collected. The actual figure will largely depend on the breeding system of the plant, the plant type and the part of the plant being collected.

Contingencies

Collecting should not take place without meeting the legal requirements, especially if the germplasm is taken out of the country of collecting afterwards. In the event that materials cannot be taken out of the country due to phytosanitary requirements, efforts should be made to establish field collections in the country of origin and/or to establish *in vitro* cultures that are more amenable for export. Allowances in terms of the sample size should be made for wild and rare species where propagation material might not be available in optimal conditions or quantity.

SELECTED REFERENCES

Alercia, A., Diulgheroff, S. & Mackay, M. 2012. FAO/Bioversity *Multi-Crop Passport Descriptors* (MCPD V.2). Rome, FAO and Bioversity International (available at: http://www.bioversityinternational.org/uploads/tx_news/1526.pdf).

Bioversity International/ Food and Fertilizer Technology Center/TARI-COA (Taiwan Agricultural Research Institute-Council of Agriculture). 2011. *A training module for the international course on the management and utilisation of field genebanks and* in vitro *collections*. Fengshan, Taiwan, TARI.

Brown, A.H.D. & Hardner, C.M. 2000. *Sampling the genepools of forest trees for ex situ conservation. In* A. Young, D. Boshier & T. Boyle. *Forest conservation genetics. Principles and practice*, pp.185–196. CSIRO and CABI.

Bustamante, P.G. & Ferreira, F.R. 2011. Accessibility and exchange of plant germplasm by EMBRAPA. *Crop Breeding and Applied Biotechnology*, S1: 95–98.

Engelmann, F., ed. 1999. *Management of field and* in vitro *germplasm collections*. Proceedings of a Consultation Meeting, 15–20 January 1996. Cali, Colombia, CIAT, and Rome, IPGRI.

FAO. 1995. Collecting woody perennials. *In* L. Guarino, V.R. Rao & R. Reid, eds. *Collecting plant genetic diversity. Technical guidelines*, pp. 485–511. Wallingford, UK, CABI.

Ferreira, F.R. & Nehra, N. 2011. *Forestry germplasm exchange and quarantine in Brazil. In* Society of American Foresters, National Convention, Honolulu, Hawaii, USA, 2-6 November 2011 (available at: http://www.eforester.org/natcon11/program/2011conventiononsitebook.pdf).

Frison, E.A. & Taher, M.M., eds. 1991. *FAO/IBPGR technical guidelines for the safe movement of citrus germplasm*. Rome, FAO and IBPGR.

Guarino, L., Rao R., V. & Reid, R., eds. 1995. *Collecting plant genetic diversity. Technical guidelines*, Wallingford, UK, CAB International.

Marshall, D.R. & Brown, A.H.D. 1975. Optimum sampling strategies in genetic resources conservation. *In* O.H. Frankel & J.G. Hawkes, eds. *Crop genetic resources for today and tomorrow*, pp. 3–80. Cambridge, UK, Cambridge University Press.

Reed, B.M., Engelmann, F., Dulloo, M.E. & Engels, J.M.M. 2004. *Technical guidelines for the management of field and* in vitro *germplasm collections*. Handbooks for Genebanks No. 7. Rome, IPGRI.

Said Saad, M. & Rao, V.R., eds. 2001. *Establishment and management of field genebank training manual*. Serdang, Malaysia, IPGRI-APO.

SGRP-CGIAR. Crop Genebank Knowledge Base. Field genebanks (available at: http://cropgenebank.sgrp.cgiar.org/index.php?option=com_content&view=article&id=97&Itemid=203&lang=english).

Veiga, R., Ares, I., Condon, F. & Ferreira, F.R. 2010. *Intercambio seguro de recursos fitogenéticos. In* Estrategia en los recursos fitogenéticos para los países del Cono Sur/IICA. pp. 75–83. Montevideo, PROCISUR, IICA.

Walter, B.M. & Cavalcanti, T.B. 2005. *Fundamentos para a coleta de germoplasma vegetal*. D.F. Brasil, Embrapa Recursos Genéticos.

5.3 Standards for establishment of field collections

Standards

5.3.1 A sufficient number of plants should be maintained to capture the genetic diversity within the accession and to ensure the safety of the accession.

5.3.2 A field genebank should have a clear map showing the exact location of each accession in the plot.

5.3.3 The appropriate cultivation practices should be followed taking into account micro-environment, planting time, rootstock, watering regime, pest, disease and weed control.

Context

It is difficult to provide specific standards for the establishment of a field genebank collection. It will depend very much on the nature of the species that are intended to be conserved. Species-specific standards will have to be developed depending on the biological characteristics of the species, its phenology, reproductive mechanism and population structure. There are three main considerations that should be taken in to account in establishing a field genebank collection: (a) how many plants per accessions should be maintained; (b) how the plants are laid out within the genebank; and (c) what cultivation practices need to be applied to ensure optimal growing conditions of the accessions in the collections.

Technical aspects

The decision for determining how many plants per accession should be planted in a field genebank hinges on the balance of the need to maintain the genetic diversity of accessions, space considerations, need for characterization and economic conditions of the field genebank. It will be different for annual and perennial plants, and whether species are seed or vegetatively propagated. In the case of seed propagated species, the sample size needs to be sufficiently large to capture the genetic diversity contained in the accession that has been collected. It is worth noting here that during collecting non-orthodox seed material, a proper sampling design that prioritizes plants for collection has to be made as it will be difficult to harbour a lot of 'within accession genetic diversity' in a field genebank collection. For vegetatively propagated species, only a small number of plants is necessary to represent the genetic diversity within the accession and to ensure the security of the accession. However, more plants may be needed in some cases, when the within-population diversity is greater than between-population diversity. Sample size can also depend on the purpose for establishing the collection, i.e. evaluation and/or distribution, which may determine different number of individuals per accession as compared to conservation purposes.

In establishing a field genebank collection, it is very important to know what accessions are being planted where. A proper planned layout and well prepared field plan will enhance efficiency of space use and management of the collection. The location of individual accessions should be clearly defined. In this respect, plot layout, design, electronic and print maps, as well as barcodes and field labels should be incorporated at the field genebank establishment phase. Considerations should be given to placing accessions in the most appropriate micro-environment in the genebank. Some plants require special environmental conditions and may need to be housed in greenhouses to have a greater environmental control (e.g. to avoid heat or cold) or require shading by other plants.

The growth habits and the adult size of the plants as well as irrigation structures and the ease of maintenance need to be considered when calculating the size of the plots. For perennial species, appropriate spacing of plants within the plot allows for proper growth of the individual plant, e.g. a tree, and avoids admixture of those crops that develop tubers on long underground stolons. In addition physical barriers should be implemented between plots to avoid admixture (gene flow), for instance by separating the plots with different species that do not cross pollinate. It helps avoid competition that may result in weak plants or could favour rapid spread of disease or insect pests. Invasive clones may require planting in cans, pots or boxes to reduce mixing or competition with less vigorous accessions. Accessions with easily distinguishable morphologies may be planted in adjacent plots when creeping, spreading or shedding of bulbils or seeds to the adjacent plot is a problem. For out-crossing species, sufficient isolation distance between plots of different accessions or measures, such as isolation cages, are required to maintain the genetic integrity of any seeds collected for distribution.

It should be emphasized that the layout and field plan are not fixed in time, and will change according to planting schedules. In the case of annuals, rotation is essential and this requires proper scheduling and additional space. It is also important to design layout so as to ensure that there are no pesticide drifts to the immediate environment.

Correctly and clearly written labels with two water resistant indelible tags are extremely important in field collections. The tags should contain information on: date, common name and field collection number. If possible, computer-produced labels should be used because they reduce transcription errors in names and numbers. Field maps (as hard copy and in digital form) are essential documents for field genebanks and provide a backup to field labels that are easily lost or destroyed. They should be developed before planting and kept updated regularly.

The establishment of field genebank collection requires that the appropriate cultivation practices, specific to the species, be adopted to ensure successful establishment of plants in the field genebank. Planting material needs to be selected carefully. Selecting only strong plants to retain in the field genebank could reduce genetic variation. The quality of initial planting material from a phytosanitary perspective is extremely important when planting new fields or replanting empty plots or when rejuvenating entire collections as long as no genetic selection would be undertaken. Only healthy material and vigorous parts of the plant should be used. Simple sanitary care, like using clean disinfected tools in the preparation of planting materials should be observed. The possibility of indexing for non-apparent diseases such as viruses and graft-transmitted pathogens (i.e. viroids, phytoplasmas and non-identified organisms) prior to establishment should be considered where possible.

Plants should be planted at the right time. Where recommendations on planting time for different species from different areas have been developed these should be followed. These should take into account optimal conditions for plant establishment, which could include temperature, moisture levels, soil type and rootstock etc. For plants propagated by grafting, one must be careful to get the rootstocks in a standardized way to do the grafting of all samples at the right time. Specific types of species are grafted on a rootstock of the same species, or a closely related one with proven good

compatibility. In those cases, the same rootstock should be used for all the accessions of that species. The rootstocks must be selected for their adaptation to soil characteristics and minimum influence on the behaviour of the grafted material. Trees should be planted on their own roots, not grafted, except if the use of rootstocks are needed to prevent disease or if the graft is the normal form of cultivation of a species.

Crops that require cross-pollination should be planted in groups by bloom date. In dioecious species, a suitable amount of male/female plants should be planted. For self incompatible species asexually propagated, the curator has to know which self-incompatibility (SI) system is presented by the species and the allelic combination in order to have a good field collection and to guarantee fruit or seeds formation. It is also important to observe the land treatment (agro-technical measures) during establishment of field collections.

Some species require additional support by planting shade trees in an appropriate design (e.g. coffee), which need to be chosen according to the local conditions and the requirements of the species. Some species grow as lianas (e.g. vanilla, many beans, cucurbits and others) and need trees, wooden sticks, wires or other installations for proper growth. It may be necessary to install very special beds for special species (mainly those from arid climates), e.g. "table beds" and shelters to keep away precipitation in certain periods of the year. The same may be true for special shading periods, irrigation or flooding times or covers to protect against frost etc. Some fruit tree species need regular pruning to express their typical appearance and remain healthy. For tree crops, another practice that should be strongly encouraged is the use of dwarfing rootstocks.

Contingencies

Some genotypes may not respond well to general propagation methods established for particular species types and research should be carried out to develop new methodologies. In the case of plants propagated with rootstock in a planting site requiring the use of a closely related species as rootstock, an interstock should be used.

It is important to consider maintaining the collection duplicated at another location (see Standards for security and safety duplication). Some genotypes, e.g. those found in forest understorey or may be disease susceptible, may not adapt well to conditions of full sun in the field and thus need to be provided with adequate shelter. This is exacerbated by resource constraints, causing a dual role for field genebanks (conservation plus crop improvement) which can lead to conflicts, e.g., in genebank layout, management, and duplication of accession. When maintaining field duplication is difficult, a possible option is to establish duplicates in the form of *in vitro* cultures.

SELECTED REFERENCES

Reed, B.M., Engelmann, F., Dulloo, M.E. & Engels, J.M.M. 2004. *Technical guidelines for the management of field and* in vitro *germplasm collections.* Handbooks for Genebanks No. 7. Rome, IPGRI.

Sebbenn, A.M. 2002. Número de árvores matrizes e conceito genéticos na coleta de sementes para reflorestamentos com espécies nativas. *Revista do Instituto Florestal de São Paulo,* V.14(2): 115–132.

SGRP-CGIAR. Crop Genebank Knowledge Base. Field genebanks (available at: http://cropgenebank. sgrp.cgiar.org/index.php?option=com_content&view=article&id=97&Itemid=203&lang=english).

5.4 Standards for field management

Standards

5.4.1 Plants and soil should be regularly monitored for pests and diseases.

5.4.2 Appropriate cultivation practices such as fertilization, irrigation, pruning, trellising, rootstock and weeding should be performed to ensure satisfactory plant growth.

5.4.3 The genetic identity of each accession should be monitored by ensuring proper isolation of accessions wherever appropriate, avoiding inter-growth of accessions, proper labelling and field maps and periodic assessment of identity using morphological or molecular techniques.

Context

Field management refers to the day-to-day curating of the field collections to ensure that plant accessions are in good health, are easily accessible and available for use. This involves many different activities including pest and disease control, proper nutrition of the plants, watering, weeding, pruning and monitoring of accessions to ensure genetic integrity of the collections.

Technical aspects

Germplasm losses due to poor health can be a major cause of genetic erosion in field genebanks. Maintaining healthy plant accessions in germplasm collections is a major challenge, especially when accessions are collected from a wide area of distribution where different pest and diseases exist. Accessions within collections can also be a source/focus of pest and disease spread if not properly managed. Therefore, it is important that strict control of plant introductions into the field genebank be exercised. In addition, current and historical levels of both insect and diseases populations must be considered. Careful inspections and recording are very important in all pest management operations. The timing of disease control is also of paramount importance since after the plant material is infected the damage is often irreversible. Modelling of climatic scenarios and diseases could also assist in the on control of new emerging pests and diseases.

Insect pests and diseases may include a very wide range of organisms depending on the target collections. Some of the most commonly associated plant germplasm pests include insects, mites, fungi, bacteria, nematodes, viruses, viroids, spiroplasma, phytoplasma, slug, snails as well as weeds. Vegetatively propagated plants may be virus-infected, leading to impairment of vigour, hardiness, and graft incompatibility, among others. During quarantine or maintenance, insect pests and diseases may be detected through a number of techniques including visual examination, isolation by agar plate method/streak plate method, moist chamber incubation, grafting, bioassays, electron microscope examination and plant diagnostic kits. The latter may include enzyme-linked immunosorbent assay (ELISA), which is easy to use, and already available for diseases of root crops (cassava, potato, beet), fruits (banana, pome, stone, and soft fruits) and vegetables. Major plant fungal and bacterial diseases must be controlled by prophylaxis, or prevention. DNA-based diagnostic kits are also extremely efficient in detecting diseases through PCR analysis of specific genes of pathogens. It is advisable to have staff trained in agronomy, horticulture, micropropagation, and pathology performing disease assessments.

Correct identification, at the time of delivery, of accessions susceptible to insect pests and diseases is desirable. It is important that field genebanks have a system in place for the identification of all associated pests and diseases for the range of crops they hold in their collection. This is especially true for those crops for which quarantined high risk pathogens have been described. Genebanks should also have procedures in place for the application of relevant diagnostic methodologies that give rigorous assurance on pest and disease status, as directed by local, regional and country requirements. In cases where a genebank does not have this capacity, these tasks should be outsourced to specialized institutions for quarantining incoming plants.

Genebank staff need to apply management practices that reduce the risks of spreading diseases within the collection. It is necessary to ensure that tools and implements, soil and footwear are properly sanitized. Integrated pest management (IPM) is a recommended approach for pest control, where possible. This program uses biological control where possible, supplementing it with pesticides and mechanical control. It can be very important to test clonal material for viruses and other graft-transmitted pathogens, as there has been much improvement in detection technology over the past decade. If unique plants are found to be infected, they should be cleaned by thermotherapy and/or tissue culture. To avoid costly therapy, it is always recommended to find similar material from "clean" or less infected sources.

The field genebank management staff must be proactive to meet the individual needs of diverse germplasm. After planting the plot, staff need to aid the growth of plants only by supplying favourable conditions for their development. Watering plants regularly during the dry season is far more important than fertilizing them. The irrigation system should be appropriate for the type of plant and the ecological conditions where the field collection is established. Fertilization of the field collection is complicated by the fact that many different types of plants are grown together. Each type of plant has special nutrition requirements due to genetic differences, size, or age. Compound mixtures can be used with low amount per plant and proper care to assure distribution. Small amounts applied at intervals may be more effective than the same total amount applied at intervals of several months. Pruning is necessary in most plants to keep their size within acceptable parameters within the plantation and in the case of trees to shape their canopy. Sometimes, only a light thinning should be made in order for the branches to have space to develop properly without excessive competition for light. This shaping and thinning operation should be entrusted to an experienced person. Due to the importance of a germplasm collection, labour must be of high quality and field maintenance should be done by trained personnel.

Competition with weeds is a much more serious problem for young plants than for old ones because of their shallower rooting system. Weed control is important for a rapid and vigorous plant growth. Weeds can be controlled by mechanic ways or using chemicals (herbicides). Herbicides can be used to reduce to a minimum the necessity of hand labour and mechanical cultivation. The type of weed control should be the recommended for each species.

In some accessions, other protection practices are needed such as frost and/or hail protection or against insect disease vectors using screenhouses. Fruit removal is also an important management practice for disease control, to avoid competition with the next year crop and to reduce stress on the plant.

In order to ensure the genetic identity of each accession, any contamination among accessions, geneflow from neighbouring plants and inter-growth of accessions should be avoided. Accessions in field collections may produce flowers and subsequent seeds that drop and could grow in the plot area. These seeds may not breed true due to heterozygosity, or may be cross-pollinated. Such involuntary seeding needs to be prohibited or rogued out. Monitoring and periodic checks should be made to ensure that each accession is properly identified and mapped in the field. Labelling is extremely important and needs to be constantly verified on site and compared to plot plans of the field genebank. Labels should be clear, concise and be as weather-proof as possible. The use of barcodes or other computer-generated labels are encouraged to reduce transcription errors. Identity of each accession should be periodically checked using morphological and molecular markers when possible (see Standards for characterization).

The maintenance practices are usually crop specific and may vary according to the intended use of the collection (conservation, evaluation, distribution). All germplasm accessions should be monitored, however frequency depends on whether the plant is herbaceous (with higher frequency of monitoring) vs. woody (less frequently monitored). All germplasm should be monitored for new animal, insect and disease pests that may be introduced into the germplasm collections. All germplasm must be monitored for vandalism as well (see Standards for security).

Contingencies

The lack of expertise in genebanks in dealing with pest and diseases can be a major limiting factor for maintaining healthy plants in the collection, for which skilled plant pathologists may be required. Genebanks should have contingency plans in place to deal with outbreaks of diseases. They should be in contact with specialized plant pathology services such as national plant pathology authorities, university laboratories or commercial laboratories, all of which may provide the services they require.

Another good practice is to rotate planting sites (where possible, especially for annually propagated species and perennials highly susceptible to soil sickness) so as to reduce the perpetuation of any soil-borne pests and diseases. Another option is to disinfect the soil. In some cases, plants can be grown in a nursery where phytosanitary conditions can be easier managed, and then be planted out in the field when plants have been acclimated.

Some accessions may be very valuable and vulnerable to pathogens. For such cases, it is important to keep them in screen houses and to keep duplicates *in vitro* or in cryopreservation as a complementary conservation backup.

Hand weeding might be required where plants might be injured by herbicide applications. Utilizing sites that do not favour pests and disease development for regeneration purposes is advisable.

SELECTED REFERENCES

Mathur, S.B. & Kongsdal, O. 2003. *Common laboratory seed health testing methods for detecting fungi.* Bassersdorf, Switzerland.

Navarro, L. 1988. Application of shoot-tip grafting *in vitro* to woody species. *Acta Horticulturae*, 227: 43–55.

Navarro, L., Civerolo, E.L., Juárez J. & Garnsey, S.M. 1991. Improving therapy methods for citrus germplasm exchange. *In* R.H. Brlansky, R.F., Lee & L.W. Timmer, eds. *Proceedings of XI Conference of the International Organization of Citrus Virologists*, pp. 400–408. Riverside, Florida, USA.

Reed, B.M., Engelmann, F., Dulloo, M.E. & Engels, J.M.M. 2004. *Technical guidelines for the management of field and* in vitro *germplasm collections.* Handbooks for Genebanks No. 7. Rome, IPGRI.

Roistacher, C.N., Navarro, L. & Murashige, T. 1976. Recovery of citrus selections free of several viruses, exocortis viroid, and Spiroplasma citri by shoot-tip grafting *in vitro. Proceedings of VII Conference of the International Organization of Citrus Virologists*, pp.186–194. Riverside, Florida, USA.

SGRP-CGIAR. Crop Genebank Knowledge Base. Field genebanks (available at: http://cropgenebank. sgrp.cgiar.org/index.php?option=com_content&view=article&id=97&Itemid=203&lang=english).

Sheppard, J.W. & Cockerell, V. 1996. *ISTA PDC handbook of method validation for the detection of seedborne pathogens.* Basserdorf, Switzerland, ISTA.

Sutherland, J.R., Diekmann, M. & Berjak, P. 2002. *Forest tree seed health.* IPGRI Technical Bulletin N° 6. Rome, IPGRI.

5.5 Standards for regeneration and propagation

Standards

5.5.1   Each accession in the field collection should be regenerated when the vigour and/or plant numbers have declined to critical levels in order to bring them to original levels and ensure the diversity and genetic integrity is maintained.

5.5.2   True-to-type healthy plant material should be used for propagation.

5.5.3   Information regarding plant regeneration cycles and procedures including the date, authenticity of accessions, labels and location maps should be properly documented and included in the genebank information system.

Context

In the context of field collections, the terms regeneration and propagation refer to the re-establishment of germplasm samples that are genetically similar to the original collection when vigour or plant numbers are low (Dulloo *et al.*, 2008). Standards for regeneration and propagation procedures would need to be species specific. When available, protocols or guidelines for particular species should be used. Regeneration and propagation should aim at ensuring that there is no loss of any plants within the collection. However, it is inevitable that the loss of any single individual would entail genetic erosion within the accession because there are normally only a few plants for each accession (see Standards for establishment of field collections: sample size). Regeneration and propagation are costly and should be carefully planned. They may require changing sites for security or to avoid diseases, pests and soil sickness processes.

Technical aspects

Regeneration and propagation may be necessary for a variety of reasons depending on the plant type, threats and distribution needs. A plant may decline in vegetative vigour or even die from many different causes, due to climatic, edaphic and/or biotic factors. For maximum efficiency in a field collection plot, it is essential that every dead plant be replaced. This is especially important since the number of individuals per accession is generally low in field collections (see Standards for establishment of field collections).

The method of propagation of the target species is an important consideration. Some species can be propagated by seeds while other species are propagated vegetatively. In principle, seeds should not be used for propagation in a field collection even if the species can reproduce by seeds unless the population size is represented by a sufficiently large number of individuals. As the objective of regeneration is to maintain the genetic integrity of the accession and, given that there is only a limited number of plants per accession, propagation through seeds can lead to significant genetic drift in the accession. In addition, in cross-pollinated species, hybridization between accessions may effectively reduce the genetic variance between accessions and change the integrity of individual accessions. Whenever possible, plants should be propagated vegetatively in which case each offspring is an exact replica of the parent and hence genetic integrity of the accession is maintained.

The time at which regeneration should be carried out is another important factor, which often depends on climate and planting season of the crop. FAO has published a series of crop calendars for Latin America and Africa (FAO, 2004, 2012), which can be a useful guide in determining the appropriate time for planting, and thus for regeneration. The FAO crop calendars provide information for more than 130 crops, located in 283 agro-ecological zones of 44 countries. Again, the timing will be species- and possibly site-specific. A good indication of when to initiate propagation is provided when propagules start to sprout or mother plants start to die continuously. Another consideration will be whether or not the collection is to be ratooned, i.e. suckers are allowed to develop to produce the next crop, as is the case for aroids (Jackson, 2008).

Propagation should be done using true-to-type and healthy plant material. If available, the new plant has to be regenerated using propagation material stored in special facilities (greenhouses, *in vitro*, or freezer) to ensure its health. Available protocols or guidelines for particular species should be used. Regeneration of accessions of out-crossing species should be made in isolation using special facilities and protection for weeds, pests and diseases.

It is important that all information relating to the regeneration of the accession be properly documented and included in the genebank documentation system. This should include *inter alia* information about the accession number and the plant sequence number within each accession, the site where regeneration is carried out, the type of propagation and materials used (cuttings, tuber, corms, bulbs), planting date, survival rate of the propagated materials, the protocol for seed dormancy-breaking, management practices employed, method of planting, field conditions, number of plants established and harvest dates).

Contingencies

Climatic factors may be more harmful to young plants than to older ones. Because a few plants are likely to be lost during the first year due to various causes, it is a wise precaution at planting to keep some plants for use as replacement if needed. This assures to have plants of the same type and age as the original for replacing lost individuals.

Field collections are extremely vulnerable to climatic and other environmental disturbances and it is very important for field genebanks to have a contingency plan for urgent regeneration of the collection. A safety backup may be maintained *in vitro* or cryopreserved as a complementary measure. Contingencies may also occur with wild relatives of crops and native species for which regeneration protocols are yet to be developed. These may often require different treatments when compared with cultivated relatives.

SELECTED REFERENCES

Costa, N., Plata, M.I. & Anderson, C. 2004. Plantas cítricas libres de enfermedades. *In* V. Echenique, C. Rubistein & L. Mroginski, eds. *Biotecnología y Mejoramiento vegetal*, pp. 317–318. Argentina, INTA.

Dulloo, M.E., Thormann, I., Jorge, A.M. & Hanson J., eds. 2008. *Crop specific regeneration guidelines.* [CD-ROM], Rome, SGRP–CGIAR.

FAO. 2004. *Calendario de cultivos. América Latina y el Caribe.* Estudio FAO producción y protección vegetal No. 186. Rome.

FAO. 2012. *Crop calendars* (available at: http://www.fao.org/agriculture/seed/cropcalendar/welcome.do).

ICRISAT (International Crops Research Institute for the Semi-Arid-Tropics). Germplasm regeneration (available at: http://www.icrisat.org/what-we-do/genebank/genebank-manual/germplasm-regeneration-9.pdf).

Jackson, G.V.H. 2008. Regeneration guidelines: major aroids. *In* M.E. Dulloo, I. Thormann, A.M. Jorge & J. Hanson eds. *Crop specific regeneration guidelines.* [CD-ROM], Rome, SGRP–CGIAR.

Plata, M.I. & Anderson, C.M. 2008. In vitro *blueberry* (Vaccinium *spp.*) *germplasm management in Argentina.* 9th International Vaccinium Symposium, ISHS, Corvallis, Oregon, USA.

Sackville Hamilton, N.R. & Chorlton, K.H. 1997. *Regeneration of accessions in seed collections: a decision guide.* J. Engels, ed. Handbooks for Genebanks No. 5. Rome, IPGRI.

SGRP-CGIAR. Crop Genebank Knowledge Base. Field genebanks (available at: http://cropgenebank. sgrp.cgiar.org/index.php?option=com_content&view=article&id=97&Itemid=203&lang=english).

5.6 Standards for characterization

Standards

5.6.1   All accessions should be characterized.

5.6.2   For each accession, a representative number of plants should be used for characterization.

5.6.3   Accessions should be characterized morphologically using internationally used descriptor lists where available. Molecular tools are also important to confirm accession identity and trueness to type.

5.6.4   Characterization is based on recording formats as provided in internationally used descriptors.

Context

Characterization is the description of plant germplasm, and a tool for the description and fingerprinting of the accessions, confirmation of their trueness to type, and identification of duplicates in a collection. It determines the expression of highly heritable characters ranging from morphological, physiological or agronomical features, including agrobotanic traits such as plant height, leaf morphology, flower colour, seed traits, phenology, and overwintering ability for perennials. These are essential information for curators to distinguish among samples in the collection.

For field collections, characterization can be carried out at any stage of the conservation process. However, it is essential that the accessions being conserved are known and described to the maximum extent possible to assure their maximum use for customers and stakeholders. Therefore, characterization should be carried out as soon as possible to add value to the collection. The time will vary from species to species depending on their life cycle. The use of a minimum set of phenotypic, physiological and morphological descriptors and information on the breeding system, selected from internationally used descriptor lists (e.g. those published by Bioversity International, the International Union for the Protection of New Varieties of Plants [UPOV] and the USDA's National Plant Germplasm System [USDA-ARS NPGS]) increases the usefulness and cross-referencing of the characterization data.

With the advances in biotechnology, molecular marker technologies and genomics are increasingly used for characterization (De Vicente *et al.*, 2004). Characterization will allow true-to-type identification, detecting gene flow and setting reference profiles, identifying mislabelling and duplications, detecting diversity within and among accessions and coefficient of parentage. Measures, such as splitting samples, may be necessary for ensuring the preservation of rare alleles or for improving access to defined alleles. Documentation of observations and measures taken is extremely important.

Technical aspects

In contrast to seed collections, phenotypic characterization of field collections is easier to perform, given that the plants are in the field and the scoring of the relevant traits for characterization can be done at the appropriate time and repeated over the years.

Some relevant characterization data can be obtained when collecting in the field, so the time for collecting expeditions should be carefully planned whenever possible. Accessions could then be characterized side by side in the field when collected. The historical and cultural information obtained from farmers, botanists, horticulturalists, or native people during collecting expeditions is usually valuable. Local knowledge about the origin of an accession and disease and insect resistance can decrease characterization costs and limit duplication.

Descriptors for crops are defined by crop experts and/or curators in consultation with crop experts and genebank managers for relevancy to increase utilization of collections. A wide range of crop descriptor lists has been developed (for example by Bioversity International, UPOV, the International Organisation for Vine and Wine (OIV) and the USDA-ARS NPGS), as well as minimum sets of key descriptors for utilization have been established for several crops. Data recording needs to be conducted by trained staff using calibrated and standardized measuring formats as indicated in the descriptor lists. The data need to be validated by curators and documentation officers before being uploaded into the genebank database and made publicly available to encourage the use of the collection. It is also recognized that reference accessions planted in the same field are needed to score the traits. Reference collections (herbarium specimens, high quality voucher images) play an essential role for true-to-type identification.

The number of plants characterized within an accession should be a representative sample, which in turn depends on its diversity. In general, there should be a minimum of 3 plants for diverse accessions, whereas for clonal plants 1-2 are sufficient[1], in order to have statistically sound measurements. In species prone to mutation (e.g. citrus), annual characterizations for key characters should be done for true-to-type verification.

With the advances in biotechnology, molecular marker technologies and genomics are increasingly used for characterization (De Vicente *et al.*, 2004), in combination with phenotypic because they have advantages on ensuring the identity of clonal plants, identifying mislabelling and duplications, detecting genetic diversity and parentages within and among accessions. Genotypic data obtained from characterizing germplasm using molecular techniques have the advantage over phenotypic data in that variations detected through the former are largely devoid of environmental influences (Bretting and Widrlechner, 1995). The technologies develop fast and costs are also decreasing quickly, allowing a more extensive use in the field collections, and should be used when resources do allow it. However, the dearth of adequately skilled personnel and the lack of resources for the relatively high set-up costs continue to prevent the widespread adoption of molecular markers as a method of choice for germplasm characterization especially in developing countries. There are many markers and techniques available (e.g. simple sequence repeats [SSR], expressed sequence tags - simple sequence repeats [EST-SSR], amplified fragment length polymorphisms [AFLP]) but, for characterization purposes, only well-established, repeatable markers such as SSR should be used.

For many crops, a wide range of marker primers suitable for their use in characterization has been developed; also, minimum sets of key markers have been established. In order to ensure that the results of different analysis batches are comparable, some genebank accessions should be included as reference on each batch. The inclusion of reference accessions in molecular characterizations also plays an essential role for comparison among different genebanks.

---

1   http://cropgenebank.sgrp.cgiar.org/index.php?option=com_content&view=article&id=47&Itemid=205&lang=english

One of the most advanced techniques employed in the improvement of tree species is genome-wide selection (GWS) (Grattapaglia and Resende 2011; Fonseca *et al.*, 2010). GWS requires the use of molecular markers that allow for wide coverage of the genome and high density genotyping. Although this technique is applied for improvement, the information generated can be used to characterize and conserve new accessions or superior genotypes.

Contingencies

Reliability of data might vary among data collectors and depends on training and experience. Therefore, trained and experienced technical staff in the field of plant genetic resources should be available during the entire growth cycle to record and document characterization data. Access to expertise in taxonomy, seed biology, plant pathology and molecular characterization (in-house or from collaborating institutes), during the process of characterization is desirable. For those crops for which there are no internationally used descriptor lists, it should be necessary to develop them while using available descriptor lists for related crops or species as references.

SELECTED REFERENCES

Alercia, A., Diulgheroff, S. & Mackay, M. 2012. FAO/Bioversity *Multi-Crop Passport Descriptors* (MCPD V.2). Rome, FAO and Bioversity International (available at: http://www.bioversityinternational.org/uploads/tx_news/1526.pdf).

Bioversity International. 2007. *Developing crop descriptor lists. Guidelines for developers.* Technical Bulletin No. 13. Rome.

Bioversity International. 2007. List of published crop descriptors (available at: http://www.bioversity international.org/index.php?id=168).

Bioversity International. 2013. Descriptor lists and derived standards (available at: http://www.bioversityinternational.org/index.php?id=168).

Bretting, P.K. & Widrlechner, M.P. 1995. Genetic markers and plant genetic resource management. *Plant Breeding Reviews*, 13:11–86.

De Vicente, M.C., Metz, T. & Alercia, A. 2004. *Descriptors for genetic markers technologies.* Rome, IPGRI.

Engels, J.M.M. & Visser, L., eds. 2003. *A guide to effective management of germplasm collections.* Handbooks for Genebanks No. 6. Rome, IPGRI.

Fang, D.Q., Roose, M.L., Krueger, R.R. & Federici, C.T. 1997. Fingerprinting trifoliate orange germplasm accessions with isozymes, RFLPs, and inter-simple sequence repeat markers. *Theor. Appl. Genet.*, 95:211–219.

Fonseca, S.M., Resende, M.D.V., Alfenas, A.C., Guimarães, L.M.S., Assis, T.F. & Grattapaglia, D. 2010. *Manual prático de melhoramento genético do eucalipto.* UFV, Viçosa, MG.

Grattapaglia, D. & Resende, M.D.V. 2011. Genomic selection in forest tree breeding. *Tree Genetics & Genomes*, 7: 241.

Lateur, M., Maggioni, L. & Lipman, E. 2010. *Report of a Working Group on Malus/Pyrus.* Third Meeting, 25-27 October 2006, Tbilisi, Georgia. Rome, Bioversity International.

Maggioni, L., Lateur, M., Balsemin, E. & Lipman, E. 2011. *Report of a Working Group on Prunus.* Eighth Meeting, 7-9 September 2010, Forlì, Italy. Rome, Bioversity International.

OIV (International Organisation of Vine and Wine). 2009. *OIV descriptor list for grape varieties and Vitis species.* 2nd ed. Paris.

SGRP-CGIAR. Crop Genebank Knowledge Base (available at: http://cropgenebank.sgrp.cgiar.org/index.php?option=com_content&view=article&id=97&Itemid=203&lang=english).

UPOV (International Union for the Protection of New Varieties of Plants). Descriptor lists (available at: http://www.upov.int/test_guidelines/en/list.jsp).

USDA, ARS, National Genetic Resources Program. Germplasm Resources Information Network - (GRIN). [Online Database] Evaluation/characterization. Data Queries. National Germplasm Resources Laboratory, Beltsville, Maryland, USA (available at: http://www.ars-grin.gov/cgi-bin/npgs/html/croplist.pl).

5.7 Standards for evaluation

Standards

5.7.1 Evaluation data on field genebank accessions should be obtained for traits of interest and in accordance with internationally used descriptor lists where available.

5.7.2 The methods/protocols, formats and measurements for evaluation should be properly documented with citations. Data storage standards should be used to guide data collection.

5.7.3 Evaluation trials should be replicated (in time and location) as appropriate and based on a sound statistical design.

Context

Evaluation is the recording of those characteristics whose expression is often influenced by environmental factors. It involves the methodical collection of data on agronomic and quality traits through appropriately designed experimental trials. Evaluation data frequently includes insect pest and disease resistance and quality evaluations (e.g. oil, protein or sugar content or density), production (wood, grain, fruits, seeds, leaf, other) and abiotic traits (drought/cold tolerance and others). These data sets are all highly desired by users to incorporate useful traits into breeding programs and help to improve utilization of the collections. The traits for which the germplasm accessions are assayed are defined in advance by crop experts in collaboration with gene bank curators. Reliable evaluation data that are easily retrievable by plant breeders and researchers facilitate greatly the use of plant germplasm accessions. Germplasm may be systematically evaluated using a network approach, at either an international level or national level.

Obtaining evaluation data by genebanks is time consuming and frequently more expensive than obtaining characterization data. Thus, evaluation should be prioritized for those accessions that have outstanding features and collaboration with breeders and other specialists (virologists, entomologists, mycologists) is recommended in this endeavour. Curators should make all possible efforts to obtain at least some minimum records of evaluation data. Possible sources of evaluation data may be obtained from users to whom germplasm materials have been distributed previously. The genebank should solicit the user to share the evaluation data and practical arrangements in this regard should be worked out between the genebank and the recipients/users of the material. Such information could address resistances to biotic and abiotic stresses, growth and development features of the germplasm, quality characteristics of yield, etc. Adding this type of information to the genebank database allows more focused identification of germplasm to meet prospective client needs. Such data should be included in the genebank's documentation system after appropriate verification and validation.

Technical aspects

A wide range of crop descriptor lists have been developed, for example, by Bioversity International and UPOV. Furthermore, there are evaluation descriptor lists developed by regional and national organizations such as USDA-ARS NPGS.

Data collection should be conducted by trained staff using as much as possible calibrated and standardized measuring formats with sufficiently identified check accessions and published crop descriptor lists. The results of greenhouse, laboratory or field evaluations, following standardized protocols and experimental procedures are usually presented as either discrete values (e.g. scores for severity of disease symptoms; counting) or continuous values (based on measuring). The data need to be validated by curators and documentation officers before being uploaded into the genebank database and made publicly available.

Many agronomic traits required by breeders are too genetically complex to be screened for in preliminary evaluations of germplasm accessions. Data on agronomic traits are usually obtained during the evaluation of germplasm in a breeding program, and many of these traits result from strong genotype by environment

(G × E) interactions and hence are site-specific. It is essential to use replications for the evaluation of desired traits in different environments and to clearly define and identify check accessions to be used over time. Check accessions facilitate comparisons across years of data collected.

With the advances in biotechnology, molecular marker technologies and genomics are increasingly used for evaluation as well (De Vicente *et al.*, 2004) (see standards on characterization). The most commonly used molecular markers in germplasm characterization and evaluation include AFLPs, SSRs, and single nucleotide polymorphisms (SNP). They have largely replaced the older marker types, restriction fragment length polymorphism (RFLP) and random amplified polymorphic DNA (RAPD) on account of their relative genomic abundance and the high reproducibility of data. In addition, advances in next generation sequencing and the accompanying reduction in costs have resulted in the increasing use of sequencing-based assays such as the sequencing of coding and non-coding regions and genotyping-by-sequencing (GBS) in germplasm evaluation. Molecular markers vary in the way they detect genetic differences, in the type of data they generate, in the taxonomic levels at which they can be most appropriately applied, and in their technical and financial requirements (Lidder and Sonnino 2011). Where marker assisted selection (MAS), i.e. the selection for the presence or absence of traits in breeding materials at the molecular level, is feasible, it can also be applied in the evaluation of germplasm for traits of interest. The dearth of adequately skilled personnel and the lack of resources for the relatively high set-up costs continue to prevent the widespread adoption of molecular markers as a method of choice for germplasm evaluation especially in developing countries.

Contingencies

Reliability of data might vary among data collectors if they are not well trained and experienced and when data collection procedures are not harmonized. Therefore, trained technical staff in the field of plant genetic resources should be available to collect and document evaluation data. The participation of multi-disciplinary teams with expertise in plant pathology, entomology, and environmental (abiotic) stress tolerance, both in-house and from collaborating institutes, during the process of evaluation is highly desirable.

The evaluation of plant germplasm is very labour-intensive and requires adequate and continuous levels of sustainable funding to allow for the assemblage

of reliable high quality data. In situations where carrying out the full evaluation of all accessions, which though desirable may not be economically feasible, the selection of genetically diverse accessions (based for instance on previously delineated sub-sets of germplasm collections) is recommended as a starting point.

Variations in the incidences of pests and diseases, the severity of abiotic stresses and the fluctuations in environmental and climatic factors in the field impact on the accuracy of data and should be mitigated through reasonably replicated, multi-locational, multi-season and multi-year evaluations. Also, the laboratory assays for the measurements of some traits like oil or protein contents, starch quality, nutritional factors, require specialized equipment and skilled staff that are not always available or could be costly. This again underscores the need for the participation of multi-disciplinary teams from several organizational units or institutions as the case may be. The sanitary status (viruses) of the accession may have incidence in the evaluation as well as in morphological descriptions.

Using the evaluation data generated by others could pose significant practical challenges. For instance, the data may be in different formats, and if published already may involve copy right and intellectual property rights issues. In order to facilitate the use of externally sourced data, it is important to standardize data collection, analysis, reporting and inputting formats.

It should be noted that many characters may appropriately be evaluated within a field-planted genebank itself. However, stresses that impose risks to the collection, and may result in accession losses if uncontrolled, should be evaluated in separate, specially designed trials. Serious insect pests and diseases or major soil problems are examples. The field collection often is not an appropriate place to evaluate yield or quality because of inappropriate plot design or the need to leave plants in the ground well beyond the normal harvest period.

SELECTED REFERENCES

Ayad, W.G., Hodgkin, T., Jaradat, A. & Rao, V.R. 1997. *Molecular genetic techniques for plant genetic resources.* Report on an IPGRI workshop, 9–11 October 1995, Rome, Italy. Rome, IPGRI.

De Vicente, M.C. & Fulton, T. 2004. *Using molecular marker technology in studies on plant genetic diversity.* Rome, IPGRI, and Ithaca, New York, USA, Institute for Genetic Diversity.

Karp, A., Kresovich, S., Bhat, K.V., Ayad, W.G. & Hodgkin, T. 1997. *Molecular tools in plant genetic resources conservation: a guide to the technologies.* IPGRI Technical Bulletin No. 2. Rome, IPGRI.

Lidder, P. & Sonnino, A. 2011. *Biotechnologies for the management of genetic resources for food and agriculture.* FAO Commission on Genetic Resources for Food and Agriculture Background Paper No. 52. Rome, FAO.

5.8 Standards for documentation

Standards

5.8.1   Passport data for all accessions should be documented using the FAO/ Bioversity multi-crop passport descriptors. In addition, accession information should also include inventory, map and plot location, regeneration, characterization, evaluation, orders, distribution data and user feedback.

5.8.2   Field management processes and cultural practices should be recorded and documented.

5.8.3   Data from 5.8.1. and 5.8.2 should be stored and changes updated in an appropriate database system and international data standards adopted.

Context

Comprehensive information about accessions, including regularly updated and detailed field maps as well as information about field management processes is essential for a field genebank to manage and maintain its field collections. Documentation of characterization and evaluation data is particularly important to enhance the use of the respective collection and to help in the identification of distinct accessions.

Technical aspects

All data and information generated throughout the process of acquisition, establishment of the collection, field management, regeneration, characterization, evaluation, and distribution should be recorded. Such data and information ranges from details of the genetic characteristics of individual accessions and populations to distribution networks, clients and user feedback. Types of data to be recorded in a field genebank other than passport data and standard crop descriptors are for example, plant catalogues, voucher images (photos, drawings), planting and harvest dates, and notes on the verification (identity) history.

The FAO/Bioversity List of Multi-crop Passport Descriptors (Alercia *et al.*, 2012) should be used for documenting passport data as they are instrumental for data exchange among different genebanks and countries. Standards for documenting characterization data such as the Bioversity International crop descriptors as well as genetic marker descriptors (De Vicente *et al.*, 2004) should be used. With advances in biotechnology, there is a need to complement phenotypic trait data with molecular data. Efforts should be made to record the molecular data being generated through genomics, proteomics, metabolomics and bioinformatics.

Record keeping about the field management processes including daily interventions, is extremely important for good management of the field collection. Good records of field maps (as hard copy and in digital form) are essential to properly document. Old maps should be retained and dated for reference.

Different cultural practices are required for the proper management of accessions of different types of species and should be carefully documented to guarantee their consistent employment over time and the appropriate treatment of accessions.

A majority of genebanks now have access to computers and the internet. Computer-based systems for storing data and information allow for comprehensive storage of all information associated with the management of field collections. Germplasm information management systems such as GRIN-Global have specifically been developed for universal genebank documentation and information management. The adoption of data standards which today exist for most aspects of genebank data management helps make the information management easier and improves use and exchange of data. Sharing accession information and making it publicly available for potential germplasm users is important to facilitate and support the use of the collection. Ultimately, conservation and usability of conserved germplasm are promoted through good data and information management.

All data should be kept up to date. They should also be duplicated at regular intervals and stored at a remote site to guard against loss from fire, computer failure etc. (see standards for security and safety). It can be useful to have written records of the main passport data and hard copies of the field maps.

Contingencies

Lack, or loss, of documentation, field plans or labels compromises the optimal use of the germplasm or can even lead to its loss, if it impedes proper management and regeneration.

Lack of adequate identification of species does not allow to record all necessary information for proper management of the accession and to identify appropriate cultural practices.

SELECTED REFERENCES

Alercia, A., Diulgheroff, S. & Mackay, M. 2012. FAO/Bioversity *Multi-Crop Passport Descriptors* (MCPD V.2). Rome, FAO and Bioversity International (available at: http://www.bioversityinternational.org/uploads/tx_news/1526.pdf).

De Vicente, M.C., Metz, T. & Alercia, A. 2004. *Descriptors for genetic markers technologies.* Rome, IPGRI.

Fabiani, A., Anderson, C. & Tillería J. 1996. *Desarrollo de una base de datos para la evaluación de germoplasma cítrico. (Abstr.).* VIII Congreso latinoamericano y VI Nacional de Horticultura. Montevideo, Sociedad Uruguaya de Horticultura

Lipman, E., Jongen, M.W.M, van Hintum, Th.J.L., Gass, T. & Maggioni L., comps. 1997. *Central crop databases: tool for plant genetic resources management.* Rome, IPGRI, and Wageningen, Netherlands, CGN.

Painting, K.A, Perry, M.C, Denning, R.A. & Ayad, W.G. 1993. *Guidebook for genetic resources documentation.* Rome, IPGRI.

Tillería, J. 2001. *Sistema DBGERMO para la Documentación de Bancos Activos de Germoplasma.* Memoria, Reunión Técnica para Latinoamérica y el Caribe del Sistema Mundial de la FAO de Información y Alerta para los Recursos Filogenéticos. pp 85–115. Turrialba, Costa Rica.

Tillería, J. & Anderson, C.M. 2004. *The DBGERMO II desktop system for an easy documentation of germplasm collections.* Proc. ISC. (Abstr.), Agadir, Morocco.

Tillería, J. & Zamuz, J. 2011. *La Herramienta Curatorial DBGERMOWeb para la Documentación de Colecciones Vegetales. Demostración de la aplicación web en tiempo real con colecciones documentadas.* VIII SIRGEALC, Quito.

Tillería, J., Andrade, R. & Zamuz, J. 2011. *Documentación de la colección de chirimoya* (Annona cherimola *Mill) del INIAP con la herramienta curatorial DBGERMOWeb.* VIII SIRGEALC, Quito.

Tillería, J., Paniego, N., Zamuz, J. & Luján, M. 2009. *El Sistema DBGERMO Web para la Documentación de Colecciones Vegetales.* VII SIRGEALC, Pucón, Chile.

USDA, ARS, Bioversity International, Global Crop Diversity Trust. GRIN-Global. Germplasm Resource Information Network Database - Version 1 (available at: http://www.grin-global.org/index.php/Main_Page).

5.9   Standards for distribution

Standards

5.9.1   All germplasm should be distributed in compliance with national laws and relevant international treaties and conventions

5.9.2   All samples should be accompanied by all relevant documents required by the donor and the recipient country.

5.9.3   Associated information should accompany any germplasm being distributed. The minimum information should include an itemized list, with accession identification, number and/or weights of samples, and key passport data.

Context

Germplasm distribution is the supply of a representative sample from a genebank accession in response to requests from germplasm users. There is a continuous increase in demand for genetic resources to meet the challenges posed by climate change, by changes in virulence spectra of major insect pests and diseases, by invasive alien species and by other end-user needs. This demand has led to wider recognition of the importance of using germplasm from genebanks, which ultimately determines the germplasm distribution. It is important that distribution of germplasm across borders adheres to international norms and standards relating to phytosanitary regulations and according to provisions of international treaties and conventions on biological diversity and plant genetic resources.

Technical aspects

The two international instruments that govern the access of genetic resources are the ITPGRFA and the CBD. The ITPGRFA facilitates access to PGRFA, and provides for the sharing of benefits arising from their utilization. It has established a multilateral system for PGRFA for a pool of 64 food and forage crops (commonly referred to as Annex 1 crops to the Treaty), which are accompanied SMTA for distribution. SMTA can also be used for non-Annex 1 crops; however, other models are also available. Access and benefit-sharing under CBD is according to its Nagoya Protocol. Both the ITPGRFA and CBD emphasize the continuum between conservation and sustainable utilization, along with facilitated access and equitable sharing of benefits arising from use.

In addition, all accessions should be accompanied with the required documentation such as phytosanitary certificates and import permits, as relevant according to the IPPC. The final destination and the latest phytosanitary import requirements for the importing country (in many countries, regulations are changed frequently) should be checked before each shipment. Germplasm transfer should be carefully planned in consultation with the national plant protection organization or the officially authorized institute, which needs to supply the appropriate documentation, such as an official phytosanitary certificate, complying with the requirements of the importing country. The recipient of the germplasm should provide the supplying genebank with information concerning the documentation required for the importation of plant material, including phytosanitary requirements.

Vegetative materials from a field genebank accession should be subjected to therapy and indexing procedures before being distributed to germplasm users. Indexing for difficult to detect pathogens, such as viruses, is important for limiting their spread. When virus indexing capabilities are unavailable, in particular for material known to have come from virus-infected areas, the sanitary status should be attached to the passport data and the material distributed if the recipient has a quarantine facilities or if it meets the criterion of the import permit of the requesting country or region.

The type of shipping container, packing materials and the choice of shipping company will depend greatly on the plant part to be distributed. Phytosanitary certificates and quarantine and import permits often document how the specific germplasm has to be packaged and shipped. Dormant or storage organs require fewer precautions and may spend a longer time in transit without damage than actively growing propagules. Accessions should be kept separate during shipment; they must not mix. Standard operating procedures (SOPS) available in many genebanks cover technical issues such as packaging, treatment, shipping method, sample size, etc. and should be referred to.

Timing shipments to avoid severe weather (either hot or cold) and notifying the recipient or customs official prior to the plant's arrival will improve the likelihood that the plants will arrive in good condition. Fragile propagules may require express delivery services. International shipments are facilitated if necessary papers are attached to the outside of the container for easy access by officials without disturbing the plants, with copies inside for the recipient. The requestor may need to purchase services of a courier to carry the germplasm through customs into the country.

All accessions should be accompanied with the minimum information necessary to the requester to make appropriate use of the material. This information should include at least an itemized list, with accession identification, number and/or weights of samples, and key passport data. In addition, the pathogen testing history is usefully included. Distribution records (records with date of request, plants requested, plant form, requester's name and address, shipment date and shipping cost) should be maintained and included in the genebank's documentation system (see standards for documentation). Distributed plant material may become a source of propagative material in case of a catastrophic loss of original material at the originator genebank.

Contingencies

Simultaneous conservation of accessions *in vitro* provides protection from pests, pathogens and climatic hazards and increases their availability for distribution if the materials are maintained virus free. In some cases, such as cassava (*Manihot esculenta* L.) and cacao (*Theobroma cacao* L.) cuttings from field banks can generally only be disseminated within a country, and sometimes only within certain regions of a country, due to pest and disease quarantine regulations. Other forms of propagation, e.g. *in vitro* cultures or seeds should be used to exchange germplasm between countries or quarantine regions. Distribution of materials from greenhouses or screenhouses may be necessary for crops with insect- or mite-borne viruses and *in vitro* cultures may be required.

Political decisions, crisis situations or bureaucratic delays might extend the time span between receipt of a sample request and the distribution of the material. Limitations related to regeneration and/or multiplication of the accessions may also affect and delay the distribution process. A delay in checking quarantine regulations until the shipment is ready to send will result in a waste of resources. Consignments of germplasm infested with pests or without proper documentation will be refused entry into the importing country or be destroyed.

SELECTED REFERENCES

SGRP-CGIAR. Crop Genebank Knowledge Base. Distribution (available at: http://cropgenebank.sgrp. cgiar.org/index.php?option=com_content&view=article&id=59&Itemid=208&lang=english).

5.10 Standards for security and safety duplication

Standards

5.10.1   A risk management strategy should be implemented and updated as required that addresses physical and biological risks identified in standards.

5.10.2   A genebank should follow the local Occupational Safety and Health (OSH) requirements and protocols.

5.10.3   A genebank should employ the requisite staff to fulfil all routine responsibilities to ensure that the genebank can acquire, conserve and distribute germplasm according to the standards.

5.10.4   Every field genebank accession should be safety duplicated at least in one more site and/or backed up by an alternative conservation method/strategy such as *in vitro* or cryopreservation where possible.

Context

Given that a field genebank is a live assemblage of plants collected from different areas that will stay in one location for many years, it is extremely vulnerable to a number of threats, including environmental conditions, pests and diseases, land tenure and land development. A field genebank is also expensive to maintain and requires constant care compared to other means of conservation. It should implement and promote systematic risk management that addresses the physical and biological risks in the every day environment. This standard provides the elements which a genebank need to fulfil in order to secure the collection for these threats and ensure that no loss in genetic diversity occurs.

Technical aspects

A field genebank should have in place a written risk management strategy on actions that need to be taken whenever an emergency occurs in the genebank concerning the germplasm or the related data. This strategy and an accompanying action plan should be regularly reviewed and updated to take advantage of changing circumstances and new technologies, and be well publicized among the genebank staff.

Field genebanks are exposed to many threats. These include extreme weather conditions like drought, freezing, hail, cyclones, typhoons, hurricanes, which are partially predictable and precautions can be undertaken to give plants additional protection during unfavourable periods. If plants are held in pots, they can be taken into a sheltered place. For smaller plants in the open field, depending on the plant type, little can be done except for reinforcing stakes or covering with a protective cover where feasible. For fruit trees, pruning branches can be done to reduce the impact of strong winds that may lead to the uprooting of trees.

Other extreme events such as fire outbreaks or earthquakes are hardly predictable and, in such cases, precautionary measures to prevent damage to plants in the field genebank need to be taken. Fire breaks across the field genebank need to be established and maintained at all times. In addition, fire-fighting equipment has to be in place and regularly checked. Fire-fighting equipment will include extinguishers and fire blankets. Field genebank buildings including greenhouses and nurseries need to be earthquake-proof if situated in a seismic-prone area.

Other threats to field collections relate to biotic factors including pests and diseases, predators, alien species, rodent pests and native material of the same species growing wild in the area that can enter the field as weeds. Precautionary measures need to be taken against these threats. Pesticides should be used with caution as this not only has a negative impact on the environment, but also on the health and safety of personnel applying these. Where appropriate, the use of traps to catch predators or ditches to prevent access to the plots can be more ecologically friendly and the invasion of animals into field genebanks should be avoided by using humane protocols approved by relevant societies.

Vandalism or theft of planting material can also be major problem to the security of collections. Field genebanks should be appropriately fenced and access to the premises of the field genebank should be well controlled. In some places, additional security guards or security fencing may be required. Considering the long-term nature of field genebanks, especially for fruit and other tree species, securing the land tenure and development plan for the site is important to reduce the need for moving to a new site and to allow expansion.

The occupational health and safety of the staff should also be considered. Properly functioning protective equipment and clothing should be provided and used in the field, especially when using chemical pesticides and fertilizers. Choice of agrochemicals is important to reduce risk. A list of chemicals that are generally safe for various crops and a "black list" of chemicals that are dangerous and are forbidden should be established. Staff should be instructed in the correct and safe use of equipment with regular training provided in health and safety in field environments.

Active genebank management requires well-trained staff, and it is crucial to allocate responsibilities to suitably competent employees. A genebank should therefore, have a plan or strategy in place for personnel, and a corresponding budget allocated regularly so as to guarantee that a minimum of properly trained personnel is available to fulfil the responsibilities of ensuring that the genebank can acquire, conserve and distribute germplasm. Access to disciplinary and technical specialists in a range of subject areas is desirable, depending on the mandate and objectives of each individual genebank. However, staff complements and training will depend on specific circumstances. Staff should have adequate training acquired through certified training and/or on-the-job training and training needs should be determined as they arise.

The use of complementary conservation methods for safety duplication of accessions maintained in field genebanks is an important strategy to reduce risk mentioned above and may be more economical. Accessions may be backed up as slow-growth *in vitro* cultures or cryo-preserved in LN, whenever protocols for the target accessions are available. For those species that produce short-lived or recalcitrant seeds, short-term seed storage, where seeds are renewed before viability is lost, is a feasible and cost effective backup method. A duplicate field genebank in another area with a suitable climate and agroecology where the plants will thrive, but that is not subject to the risks of the main genebank, can also be used for safety backup. It also provides an additional site from which material can be distributed and can be located in an area with different pest and disease risks for safety of the collection and to ease quarantine restrictions for distribution within regions. Pollen and DNA storage also complements field genebanks by providing a cost effective way to maintain a larger amount of diversity within an accession than can be maintained as plants in the field genebank.

Any safety duplication arrangement requires a clear signed legal agreement between the depositor and the recipient of the safety duplicate that sets out the responsibilities of the parties and terms and conditions under which the material is maintained. This is particularly important for field genebanks where the plants have to be managed on a daily basis.

Contingencies

When suitably trained staff is not available, or when there are time or other constraints, the solutions would be to include outsourcing some of the genebank work or approaching other genebanks for assistance. It is important to develop networks and collaborations with other genebanks. The international community of genebanks should be instantly informed, if the functions of the genebank are endangered.

Unauthorized entry to genebank facilities by humans or incursion of animals, including birds and other wildlife can result in direct loss of material, but can also jeopardize the collections through inadvertent introduction of insect pests and diseases and interference in management systems. Working closely with local communities to raise awareness of the purpose and value of the collection can give a sense of ownership and increased protection to the field area.

SELECTED REFERENCES

Engels, J.M.M. & Visser, L., eds. 2003. *A guide to effective management of germplasm collections*. Handbooks for Genebanks No. 6. Rome, IPGRI.

NordGen. 2008. *Agreement between (depositor) and the Royal Norwegian Ministry of Agriculture and Food concerning the deposit of seeds in the Svalbard Global Seed Vault*. The Svalbard Global Seed Vault. The Nordic Genetic Resource Centre, ALNARP (available at: http://www.nordgen.org/sgsv/scope/sgsv/files/SGSV_Deposit_Agreement.pdf).

Reed, B.M., Engelmann, F., Dulloo, M.E. & Engels, J.M.M. 2004. *Technical guidelines for the management of field and* in vitro *germplasm collections*. Handbooks for Genebanks No. 7. Rome, IPGRI.

SGRP-CGIAR. Crop Genebank Knowledge Base. Safety duplication (available at: http://cropgenebank.sgrp.cgiar.org/index.php?option=com_ content&view=article&id=58&Itemid=207&lang=english).

6

Genebank standards for *in vitro* culture and cryopreservation

The Standards for *in vitro* culture and cryopreservation are broad and generic in nature due to the marked variation among non-orthodox seeds and vegetatively propagated plants. This variability is a function of the inherent biology and metabolic status of the plants concerned, which influences their differing responses to various manipulations and often requires modifications of basic approaches to be made on a species-specific basis. These various features necessitate an introduction to the phenomenon of non-orthodoxy and storage behaviour of non-orthodox seeds to better understand the scientific basis of these standards.

Phenomenon of non-orthodoxy

Understanding the desiccation tolerance and sensitivity in orthodox compared to non–orthodox (intermediate and recalcitrant) seeds is of fundamental importance for cryopreservation. At maturity, orthodox seed water content would generally be in the range 0.05 – 0.16 g g^{-1} [1] (5 percent - 14 percent [wmb]), although some species are shed at much higher water content, undergoing substantial dehydration after this. Unlike recalcitrant seeds, all orthodox seeds acquire desiccation tolerance, which is genetically-programmed and entrained before, or at the start of maturation drying. Recalcitrant seeds do not dry during the later stages of development and

---

1   In this document, the term water content (wmb here is wet mass basis) is used in preference to moisture content, as recalcitrant seeds are hydrated (wet) rather than moist (barely wet). Also, the figures given are expressed on a dry mass basis (g H$_2$O g^{-1} dry matter [g g^{-1}]), which is considered to be more explicit than expression as a percentage of the wet mass.

are shed at water contents in the range of 0.3–0.4 – >4.0 g g^{-1}. Because they are desiccation-sensitive, the loss of water rapidly results in decreased vigour and viability, and seed death at relatively high water contents. This is due to their metabolic activity (Berjak and Pammenter, 2004) with little or no intracellular differentiation occurring, thus exposing membranes to the damaging consequences of dehydration stress (Walters *et al.*, 2001; Varghese *et al.*, 2011). A spectrum of differences in post-shedding physiology also occurs in intermediate seeds. Seeds showing intermediate behaviour can withstand water loss to ~ 0.11 to ~ 0.14 g g^{-1} (Berjak and Pammenter, 2004). They have the capacity to perform some of the important mechanisms and processes governing desiccation tolerance. However, they are not long-lived in the dehydrated condition, particularly at chilling temperatures for some species.

The variability in physiology of recalcitrant seeds is frequently also intraspecific. Seed or, embryo/embryonic axis water content can vary significantly in collections from the same locality from year-to-year, and also for material from the same locality, within any one season. This means that the parameters (water content, response to drying) must be assessed for each species. Additionally, seeds harvested late in a season are usually of considerably inferior quality compared with those harvested earlier (Berjak and Pammenter, 2004). The provenance of the population from which seeds are collected is also a major factor in the properties and responses of recalcitrant seeds. Thus, even if they are of the same species, seeds developing along a latitudinal gradient can show remarkably different characteristics. (Daws *et al.*, 2006; Daws *et al.*, 2004).

Seed developmental status has emerged as a critical consideration when recalcitrant germplasm is to be cryostored. Early during seed ontogeny, all seeds are highly desiccation-sensitive. Desiccation sensitivity in recalcitrant seeds increases as the processes of germinative metabolism are manifested (Berjak and Pammenter, 2004). The early events of germination in recalcitrant seeds are initiated soon after they are shed, without the 'punctuation' between the end of development and the start of germination imposed on orthodox seeds by maturation drying.

Depending on the species, recalcitrant seeds will initiate germinative metabolism after being shed. Those species with fully developed embryos on shedding, generally initiate germination virtually immediately, with a concomitant increase in desiccation sensitivity. In some other species, seeds are shed with under-developed embryos, necessitating the completion of development prior to the onset of germinative metabolism. These developmental differences dictate the duration for which the seeds can be wet-stored (i.e. hydrated storage at the shedding water content). It is now known that recalcitrant seeds cannot be dehydrated to a water content precluding germination (so-called sub-imbibed storage), as this actually shortens the hydrated storage life span. Slight dehydration actually stimulates the onset/progression of germination, thus shortening the time before an extraneous water supply is required to support the process (Drew *et al.*, 2000; Eggers *et al.*, 2007).

In general, recalcitrant seeds from temperate provenances are chilling-tolerant, while those from the tropics and sub-tropics and of the same species, are more likely to be chilling-sensitive. Chilling sensitivity is also an issue for the storage of intermediate seeds, particularly those from the tropics and sub-tropics. When dried to water contents that are not injurious in themselves, the storage life span of such seeds is curtailed at temperatures ≤10 °C (Hong *et al.*, 1996).

Seed-associated microflora (fungi and bacteria), especially those associated with the interior surfaces, e.g. of the cotyledons or embryonic axis, is generally a major problem with recalcitrant seeds, particularly of tropical and sub-tropical origin (Sutherland *et al.*, 2002). The conditions of hydrated storage, being moist and often necessarily at benign temperatures, encourage fungal proliferation, with the probability of hyphae penetrating the embryo tissues. This has a major deleterious effect and curtails hydrated storage life span significantly.

Under field conditions, unless seedling establishment is rapid, recalcitrant seeds will gradually lose water, the rate depending on the species-specific nature and morphology. Under conditions of slow water loss (days to a week or more), desiccation damage accumulates and the seeds of most species will have lost viability when the embryos/embryonic axes are at a water content of around 0.8 g g^{-1} (Pammenter *et al.*, 1993). Thus when handling or storing recalcitrant seeds, great care is normally exercised to maintain water contents at the levels characteristic of shedding.

The response of explants to dehydration depends on the rate of drying and size of the explants. Often recalcitrant seeds are too large to dry rapidly, and too large to cool rapidly on exposure to cryogen (as is required to obtain successful cryopreservation). Thus, excised embryos or embryonic axes are explants of choice, since they can be dehydrated to water contents that will minimize ice crystallization, which are $\leq 0.4 \text{ g g}^{-1}$. Embryos/axes can be dried in a stream of air (flash-dried) (Pammenter *et al.*, 2002), which significantly curtails the time during which metabolism-linked desiccation damage can occur. It is not that the embryos/axes have become desiccation-tolerant, but simply that they dry before lethal damage has accumulated, providing the time needed to subject them to cryogenic temperatures. In cases where embryo/axes prove impossible to manipulate for successful cryostorage, alternative explants, such as shoot apical meristems excised from seedlings developed from seeds germinated *in vitro*, can be used.

In addition to cryopreservation, other means of conservation for species producing recalcitrant or otherwise non-orthodox seeds include *in vitro* conservation that could involve slow growth of seedlings/young plants/plantlets. In some instances, slow-growth conditions may be imposed *ex vitro*. In the last instance, plantlets may be derived from embryogenic callus (which itself might be amenable to cryopreservation) and conserved *in vitro*, possibly under slow-growth conditions.

SELECTED REFERENCES

Benson, E.E., Harding, K., Debouck, D., Dumet, D., Escobar, R., Mafla, G., Panis, B., Panta, A., Tay, D., Van den Houwe, I. & Roux, N. 2011. *Refinement and standardization of storage procedures for clonal crops - Global Public Goods Phase 2: Part I. Project landscape and general status of clonal crop* in vitro *conservation technologies*. Rome, SGRP-CGIAR.

Berjak, P. & Pammenter, N.W. 2004. Recalcitrant Seeds. *In* R.L. Benech-Arnold, & R.A. Sánchez, eds. *Handbook of seed physiology: applications to agriculture*, pp. 305–345. New York, USA, Haworth Press.

Daws, M.I., Lydall, E., Chmielarz, P., Leprince, O., Matthews, S., Thanos, C.A. & Pritchard, H.W. 2004. Developmental heat sum influences recalcitrant seed traits in *Aesculus hippocastanum* across Europe. New Phytologist, 162: 157–166.

Daws, M.I., Cleland, H., Chmielarz, P., Gorian, F., Leprince, O., Mullins, C.E., Thanos, C.A., Vandvik, V. & Pritchard, H.W. 2006. Variable desiccation tolerance in *Acer pseudoplatanus* seeds in relation todevelopmental conditions: a case of phenotypic recalcitrance? *Functional Plant Biology*, 33: 59–66.

Drew, P.J., Pammenter, N.W. & Berjak, P. 2000. 'Sub-imbibed' storage is not an option for extending longevity of recalcitrant seeds of the tropical species, *Trichilia dregeana* Sond. *Seed Science Research*, 10: 355–363.

Eggers, S., Erdey, D., Pammenter, N.W. & Berjak, P. 2007. Storage and germination responses of recalcitrant seeds subjected to mild dehydration. pp. 85–92. *In* S.Adkins, S. Ashmore, S.C. Navie, eds., *Seeds: biology, development and ecology*. Wallingford, UK, CABI.

Engelmann, F. & Takagi, H., eds. 2000. *Cryopreservation of tropical plant germplasm. Current research progress and application*. Tsukuba, Japan, Japan International Research Centre for Agricultural Sciences, and Rome, IPGRI.

Hong, T.D., Linington, S. & Ellis, R.H. 1996. *Seed storage behaviour: A compendium*. Handbooks for genebanks No. 4. Rome, IPGRI.

Lync P., Souch, G., Trigwell, S., Keller, J & Harding, K. 2011. Plant cryopreservation: from laboratory to genebank. *As. Pac J. Mol. Biol. Biotechnol.*, 18 (1): 239–242.

Pammenter, N.W., Vertucci, C. & Berjak, P. 1993. Responses to dehydration in relation to non-freezable water in desiccation-sensitive and -tolerant seeds. *In* D. Côme & F. Corbineau, eds. *Proceedings of the Fourth International Workshop on Seeds: Basic and Applied Aspects of Seed Biology*, pp.867–872. Angers, France. ASFIS, Paris. Vol. 3.

Pammenter, N.W., Berjak, P., Wesley-Smith, J. & Willigen, C.V. 2002. Experimental aspects of drying and recovery. *In* M. Black & H.W. Pritchard, eds. *Desiccation and survival in plants: drying without dying*, pp. 93–110. Wallingford, UK, CABI.

Reed, B.M. 2010. *Plant cryopreservation. A practical guide*. New York, USA, Springer.

Reed, B.M., Engelmann, F., Dulloo, M.E. & Engels, J.M.M. 2004. *Technical guidelines for the management of field and* in vitro *germplasm collections*. Handbooks for Genebanks No. 7. Rome, IPGRI.

Sutherland, J.R., Diekmann, M. & Berjak, P., eds. 2002. *Forest tree seed health*. IPGRI Technical Bulletin No. 6. Rome, IPGRI.

Varghese, B., Sershen, Berjak, P., Varghese, D. & Pammenter, N.W. 2011. Differential drying rates of recalcitrant *Trichilia dregeana* embryonic axes: A study of survival and oxidative stress metabolism. *Physiologia Plantarum*, 142: 326–338.

Walters, C., Pammenter, N.W., Berjak, P. & Crane, J. 2001. Desiccation damage, accelerated ageing and respiration in desiccation-tolerant and sensitive seeds. *Seed Science Research*, 11: 135–148.

6.1 Standards for acquisition of germplasm

Standards

6.1.1 All germplasm accessions added to the genebank should be legally acquired, with relevant technical documentation.

6.1.2 All material should be accompanied by at least a minimum of associated data as detailed in the FAO/Bioversity multi-crop passport descriptors.

6.1.3 Only material in good condition and of consistent maturity status should be collected, and the sample size should be large enough to make genebanking a viable proposition.

6.1.4 The material should be transported to the genebank in the shortest possible time and in the best possible conditions.

6.1.5 All incoming material should be treated by a surface disinfectant agent to remove all adherent microorganisms and handled so that its physiological status is not altered, in a designated area for reception.

Context

Acquisition is the process of collecting or requesting germplasm (seeds and other propagules[1]) for inclusion in the genebank, together with related information. Adherence to legal requirements is essential, and both national and international

---

1 In this context, a propagule refers to vegetative portions of a plant such as seeds, buds, corms, cuttings and other offshoots, used to propagate a plant.

requirements must be fulfilled as appropriate. During the acquisition phase, it is important to ensure that passport data for each accession is as complete as possible and fully documented (Alercia *et al.*, 2001).

There is a need to ensure maximum quality of the germplasm and avoid conservation of immature seeds and seeds that have been exposed for too long to the elements. The way that seeds and other propagules are handled after collection and before they are transferred to controlled conditions is critical for quality. Unfavourable extreme temperatures and humidity during the post-collecting period and during transport to the genebank could cause rapid loss in viability and reduce longevity during storage. The same applies to post-harvest handling within the genebank. The seed quality and longevity is affected by the conditions experienced prior to storage within the genebank. As recalcitrant seeds are metabolically active and have high water contents at maturity, the way they are handled after collection is critical for successful long-term conservation of the material. As field-grown material is frequently contaminated with fungi and/or bacteria, it is necessary to have a set of measures in place to reduce the risk of deterioration of the material in the post-harvest state.

Material must be as clean as possible. Therefore, transfer of field material into pots and short periods of glasshouse growth is recommended. In these cases, plants should be watered from the bottom and, in extremely infected material, pesticides may support later disinfection of the explants. Visibly infected material should be excluded from the beginning or eliminated when found.

Technical aspects

Plant genetic resources within the Multilateral System of the ITPGRFA are accompanied by a Standard Materials Transfer Agreement (SMTA). For material acquired or collected outside the country where the genebank is located, the acquirers should comply with the relevant national and international legislation. Phytosanitary regulations and any other import requirements must be sought from the relevant national authority of the receiving country.

Passport data is needed to identify and classify the accessions. Many accessions are wild species, making collection of accurate field data absolutely imperative. The multi-crop passport descriptors should therefore, include a herbarium voucher, as well as GPS coordinates and photographic images of habit, habitat, and the substratum as much as possible. If fallen material is collected, it should, be recorded as such and kept separate from that harvested from the parent plant. The sample size should include an adequate number of individuals/accessions, large enough to establish an appropriate protocol for cryopreservation, and/or to place samples in long-term cryostorage.

There is a need to ensure maximum seed and propagule quality and avoid conservation of immature or over-ripe material (in the case of seeds) that has been exposed for too long to the elements. Collecting well-matured clean and high quality propagules, will ensure maximum longevity in storage. Fallen material and fruits (seeds) showing abrasions or signs of weathering should be avoided. Late-season seeds appear often to be of inferior quality to those produced earlier (Berjak and Pammenter, 2004). It is advised not to collect late-season recalcitrant seeds of any species. Seasonality needs also to be considered when using bulbs and tubers, which develop new shoots only in some seasons, in woody plants that have dormant buds only in winter, and young inflorescence explants or pollens which are available only in the flowering period.

Many of the fruits bearing recalcitrant seeds harbour fungal contaminants, even when not visible. This is a serious problem, and surface disinfection prior to transport is important for removal of any superficial contaminant. High temperatures and humidity during the post-collecting period and during transport to the genebank exacerbate this problem and could cause rapid loss in viability and reduce longevity during storage. However, seeds and other propagules may be chilling-sensitive and elevated temperatures may either hasten germination or damage the seeds. Thus, transport temperature must be neither too low nor too high, generally not below ~ 16 °C and not above ~ 25 °C.

The problem of fungal contamination also applies to post-harvest handling within the genebank and fruits should be thoroughly surface-disinfected prior to opening. Similarly, for any imported accessions, contamination can result from containers and wrappings, which need to be incinerated as is generally stipulated by national Plant and Seed Health regulations. Fruit pulp, fibres, etc. must be completely removed from seed outer surfaces, but water must not be used, as seeds could well become (further) hydrated and affect the water content of the seeds. It is also important to gather information about the fruit and seeds weight prior to water content determination (see Standard 6.2).

Wherever possible (as in the case of hard-coated fruits), seeds should be transported within the fruits, both for protection and to avoid dehydration. Water loss both stimulates germinative metabolism and shortens storage life span, thus it is important that water content are maintained upon collection and during transport, by maintaining high relative humidity (RH) in the storage containers. Special plastic bags should be preferred, which are not vulnerable to breaking as are glass tubes.

Insulating packaging will help in keeping the temperature stabile, and can be especially relevant during long transportation.

Recalcitrant seeds produced in hard-coated fruits generally remain in better condition for longer periods, than if the seeds are removed from the fruits. Soft fruits, or those which are damaged or have dehisced should immediately be surface-decontaminated, the seeds extracted and the fruits removed and destroyed. If long transport periods are involved, it is advisable to extract, manually clean and surface-disinfect the seeds prior to transport. Ideally, a disinfection kit comprising water purification tablets or sodium hypochlorite (NaOCl), water (sterile, if possible, or boiled on site) and sterile paper towelling should be carried on field expeditions.

Under tropical conditions, other measures such as the storage of plantlets under shade (Marzalina and Krishnapillay, 1999) or *in vitro* field collecting (Pence *et al.*, 2002; Pence and Engelmann, 2011) may be applied. Minimum transportation times are also necessary when *in vitro*-collected material is used.

For *in vitro*-cultured explants, surface decontamination starts often by 70 percent ethanol followed by NaOCl diluted from pure stock solution or as constituent of a commercial bleach with a concentration of active chlorine amounting to about 3 percent. Detergent droplets may support the effect. Other substances may be used as well (e.g. calcium hypochlorite) in appropriate concentrations. The explant needs to be trimmed to the final size after surface disinfection. Note that the disinfectant will enter cut surfaces resulting in dead zones that need to be removed upon trimming.

Contingencies

When a consignment is contaminated or deteriorated, all material and its packaging must be incinerated, despite the financial implications.

Delays of a consignment in national quarantine facilities are a known hazard. In such cases, steps must be taken to minimize such delays, including the use of couriers.

Under conditions of a 'poor' fruiting season, it is preferable to postpone collection to a subsequent fruiting season. If circumstances dictate that fallen fruits have to be collected, only those that are newly-abscised, should be considered.

Occasionally, seeds of particular species react badly to NaOCl and/or to the commonly used fungicides, in which case safe alternatives must be used (Sutherland *et al.*, 2002). Note that 70 percent (v/v) ethanol in sterile/boiled water could be used.

SELECTED REFERENCES

Alercia, A., Diulgheroff, S. & Mackay, M. 2012. FAO/Bioversity *Multi-Crop Passport Descriptors* (MCPD V.2). Rome, FAO and Bioversity International (available at: http://www.bioversityinternational.org/uploads/tx_news/1526.pdf).

Berjak, P. & Pammenter, N.W. 2004. Recalcitrant seeds. *In* R.L. Benech-Arnold & R.A. Sánchez, eds. *Handbook of seed physiology: applications to agriculture*, pp. 305–345. New York, USA, Haworth Press.

Engelmann, F. 1997. *In vitro* conservation methods. *In* J.A. Callow, B.V. Ford-Lloyd & H.J. Newbury, eds. *Biotechnology and plant genetic resources*, pp. 119–161. Wallingford, UK, CABI.

ENSCONET (European Native Seed Conservation Network). 2009. *Seed collecting manual for wild species* (avaliable at www.ensconet.eu).

Marzalina, M. & Krishnapillay, B. 1999. Recalcitrant seed biotechnology applications to rainforest conservation. *In* E.E. Benson, ed. *Plant conservation biotechnology*, pp. 265–276. London, Taylor & Francis.

Pence, V.C. 1996. *In vitro* collection (IVC) method. *In* M.N. Normah, M.K. Narimah & M.M. Clyde, eds. In vitro *conservation of plant genetic resources*, pp. 181–190. Percetakan Watan Sdn. Bdh, Kuala Lumpur.

Pence, V.C. & Engelmann, F. 2011. Collecting *in vitro* for genetic resources conservation. *In* L. Guarino, V. R. Rao & E. Goldberg. *Collecting plant genetic diversity: technical guidelines*. 2011 update. Rome, Bioversity International.

Pence, V. C., Sandoval, J., Villalobos, V. & Engelmann, F., eds. 2002. In vitro *collecting techniques for germplasm conservation*. IPGRI Technical Bulletin No. 7. Rome, IPGRI.

Sutherland, J.R., Diekmann, M. & Berjak, P., eds. 2002. *Forest tree seed health*. IPGRI Technical Bulletin No. 6. Rome, IPGRI.

6.2 Standards for testing for non-orthodox behaviour and assessment of water content, vigour and viability

Standards

6.2.1   The storage category of the seed should be determined immediately by assessing its response to dehydration.

6.2.2   The water content should be determined individually, on separate components of the propagule, and in a sufficient number of plants.

6.2.3   The vigour and viability should be assessed by means of germination tests and in a sufficient number of individuals.

6.2.4   During experimentation, cleaned seed samples should be stored under conditions that do not allow any dehydration or hydration.

Context

Maintaining seed viability is a critical genebank function that ensures germplasm is available to users and is genetically representative of the population from which it was acquired. As a first step to preservation, it is important to ascertain the seed storage category by assessing the response of the propagule to dehydration. The response to drying will in turn determine the treatment needed for cryostorage. A number of factors influence drying rate, including RH, seed size, the nature of seed coverings, the flow rate of air over the seeds, and the depth of the layer of seeds (Pammenter *et al.*, 2002).

The rate and uniformity of germination of a seed sample, or of seed-derived explants, is a reliable indicator of vigour, while the totality of germination (i.e. what proportion/percentage of seeds or explants tested finally germinated) reveals the overall viability of the sample. Viability should not be not less than 80 percent in a sample.

Technical aspects

Water content determinations and assessment of vigour and viability should be carried out as a single operation, and are issues to determine before selecting the type of drying technique. The number of procedures that can be investigated is determined by the number of seeds available. Three methods for screening seeds can be used for seed categorization. These includes a method that can discriminate between intermediate and recalcitrant seeds (Hong and Ellis, 1996), one which is designed for cases when seeds are limited (Pritchard *et al.*, 2004), and one that assesses axis water content, rather than that of whole seeds. Irrespective of the method chosen, dehydration imposed during the screening procedure must never be carried out at elevated temperatures, which are damaging. The recommended temperatures for tropical/sub-tropical species and those of temperate origin are 25 °C and 15 °C respectively (Pritchard *et al.*, 2004). A drying time-course assessing loss of viability with declining water content should be determined for each new accession.

The water content present within different components of recalcitrant seeds is critical for their successful cryopreservation. Water content determined on a whole recalcitrant seed basis, gives no indication of the water content of the axis. Therefore, water content determinations must be carried out separately for axes, embryos, fleshy cotyledonary tissue or endosperm (Berjak and Pammenter, 2004) and measured individually (not on pooled samples). In many cases, the dry mass of axes of recalcitrant seeds may be as little as a few milligrams, necessitating a 6-place balance.

It is important to determine water content of each newly arrived accession immediately after the propagules have been cleaned, to avoid further drying. Even if other accessions of the same species have been collected, one cannot assume that water contents will be similar. Because the composition of the axes and storage tissues of recalcitrant or otherwise non-orthodox wild species is generally unknown, drying is recommended to be carried out at 80 °C until constant weight is attained. When tissues are dried at 80 °C, the time taken to attain constant weight is generally between 24 and 48 h. After the drying period, it is imperative that samples reach room temperature, without absorbing water, before being re-weighed.

A minimum of 10 seeds is recommended to be tested for water content (determined on an individual seed/embryo/axis basis). Additional seeds will be required for any biochemical analyses undertaken.

Seeds and the embryos/axes excised from them should be at their most vigorous stage of development when newly harvested. Intact seeds are best germinated on 0.8 – 1 percent water agar in closed plastic containers or Petri dishes, which will provide common conditions for all such assessments. It is important that the surfaces of seeds are disinfected prior to being set to germinate, or prior to excision of embryos or embryonic axes. Dormancy is not a common feature of recalcitrant seeds, and seeds should normally commence germination in a relatively short interval after being set out. However, the time will vary among species depending on the extent of embryo development. It is essential that all germination/viability testing is done under the same controlled conditions per species. Production of morphologically abnormal seedlings/plantlets (Pammenter *et al.*, 2011) should be noted and quantified, as abnormality can occur as a result of imposed stress (e.g. dehydration of recalcitrant seeds, embryos or embryonic axes). A minimum of 20 seeds is recommended for viability testing.

When handling recalcitrant seeds, great care is normally exercised to maintain water contents at the levels characteristic of shedding. However, intact recalcitrant seeds are almost invariably too large to be cooled to cryogenic temperatures. Hence, the explants, embryos or the embryonic axes, need to be excised from the seeds and dehydrated. Further to this, it is essential that the bulk of the cleaned seed sample be stored under conditions that preclude changes in water status. If exposed to the atmosphere for any length of time, the water content of seeds will change and seeds shedding at relatively high water contents would become somewhat dehydrated.

Contingencies

If a genebank does not have a temperature- and humidity-controlled drying room then, for whole seeds, bench-top drying in bell jars or drying in the shade in monolayers could be used. Specimens in any Petri dish not closed before extraction from the drying oven, will have to be replaced in the oven, as dry tissues rapidly adsorb water vapour, especially in a humid environment.

Excised embryos/embryonic axes will generally not germinate as rapidly as will the intact seeds. When working with excised embryonic axes, often shoot development will not occur. In such cases, root production will be the criterion on which vigour and viability are assessed.

In cases where embryos/axes prove impossible to manipulate for successful cryostorage, alternative explants must be used. These can be of a variety of types, but the most suitable are shoot apical meristems excised from seedlings developed from seeds germinated *in vitro*.

SELECTED REFERENCES

Berjak, P. & Pammenter, N.W. 2004. Recalcitrant seeds. *In* R.L. Benech-Arnold & R.A. Sánchez, eds. *Handbook of seed physiology: applications to agriculture*, pp. 305–345. New York, USA, Haworth Press.

Hong, T.D. & Ellis, R.H. 1996. *A protocol to determine seed storage behaviour*. IPGRI Technical Bulletin No. 1. Rome, IBPGR.

Pammenter, N.W., Berjak, P., Wesley-Smith, J. & Willigen, C.V. 2002. Experimental aspects of drying and recovery. *In* M. Black & H.W. Pritchard, eds. *Desiccation and survival in plants: drying without dying*, pp. 93–110. Wallingford, UK, CABI.

Pammenter, N.W., Berjak, P., Goveia, M., Sershen, Kioko, J.I., Whitaker, C. & Beckett, R.P. 2011. Topography determines the impact of reactive oxygen species on shoot apical meristems of recalcitrant embryos of tropical species during processing for cryopreservation. *Acta Horticulturae*, 908: 83–92.

Pritchard, H.W., Wood, C.B., Hodges, S. & Vautier, H.J. 2004. 100-seed test for desiccation tolerance and germination: a case study on eight tropical palm species. *Seed Science and Technology*, 32: 393–403.

6.3 Standards for hydrated storage of recalcitrant seeds

Standards

6.3.1   Hydrated storage should be carried out under saturated RH conditions, and seeds should be maintained in airtight containers, at the lowest temperature that they will tolerate without damage.

6.3.2   All seeds should be disinfected prior to hydrated storage and infected material should be eliminated.

6.3.3   Stored seeds must be inspected and sampled periodically to check if any fungal or bacterial contamination has occurred, and whether there has been any decline in water content and/or vigour and viability.

Context

For provision of planting stock for re-introductions and restoration programmes, or simply for the maintenance of seeds whilst undertaking experimentation, it is sometimes necessary to store recalcitrant seeds in the short- to medium-term (weeks to months). The basic principle for maximizing the storage life span of recalcitrant seeds is that water contents should be retained at essentially the same levels as those characterizing the newly harvested state. Thus, the seeds must not lose water either before or after being placed in storage. Even very slight degrees of dehydration can stimulate the initiation of germination, and further dehydration can initiate deleterious changes that impact on vigour and viability and shorten the period for which the seeds can be stored. Keeping recalcitrant seeds under conditions that will maintain their water content is termed, hydrated storage, and is achieved by holding the seeds in closed conditions under saturated RH.

Technical aspects

To avoid any water loss from the seeds, hydrated storage must be carried out at saturated RH, achieved by maintaining a saturated atmosphere in the storage containers. Ideally, sealing polythene bags with an inner paper bag inside ('bag within a bag') or sealing plastic buckets of appropriate size for the seed numbers, are favoured for storage (Pasquini *et al.*, 2011). As an essential precaution, storage containers such as buckets with sealing lids, as well as internal grids, must be sterilised prior to the introduction of seeds. Irrespective of the container chosen, a means for absorbing any condensate needs to be included, and changed on becoming damp.

Storage temperatures should be the lowest that seeds of individual species will tolerate, without any deleterious effect on vigour and viability. This will slow both progress towards germination and fungal proliferation. The temperature of the store must be kept constant to minimize condensation on the interior surfaces of the storage containers. For recalcitrant seeds of temperate origin, temperatures of 6 ± 2 °C are generally suitable for storage, while for the majority of seeds of tropical/sub-tropical origin, 16 ± 2 °C is normal range. Exceptions occur, particularly for seeds of some equatorial species (Sacandé *et al.*, 2004; Pritchard *et al.*, 2004).

Under hydrated storage conditions, fungi (or less frequently bacteria) are likely to proliferate, so vigilance and appropriate action to curtail seed-to-seed infection is required. If infected seeds are not removed, they will contaminate the entire batch in a storage container. This renders the stored seeds useless and eliminates their potential for supplying explants for cryopreservation. Hence, regular inspection right from the outset, and appropriate action such as the application of fungicidal agents should be done to eliminate surface and internal contaminants from seeds at the earliest possible opportunity (Calistru *et al.*, 2000).

Seed surfaces need to be disinfected, dried of any residual sterilant, and dusted with a broad-spectrum fungicide. Internally-borne fungi, largely located immediately below the seed coverings, would be most effectively eliminated by the uptake of appropriate systemic fungicides by the seeds. However, these may well affect the seeds adversely. Thermotherapy, as applied to infected acorns (Sutherland *et al.*, 2002), is another possibility, but this can be used only when seeds are resilient to transitorily-raised temperatures – which is not always the case. To disinfect inner surfaces directly, it is necessary to ensure that the seeds survive well in hydrated storage after removal of the coverings, and that presence of systemic fungicides in the seed tissues are not damaging.

Depending on the duration of hydrated storage, containers should be briefly and periodically ventilated to avoid development of anoxic conditions at which time the contents of containers must be inspected and any contaminated seeds discarded. Storing seeds in a monolayer is ideal, but if seeds are stored in several layers, the seeds should be mixed about during aeration. After removing any seed showing signs of contamination, the container must be emptied, all apparently uncontaminated seeds disinfected and the seed lot replaced in a sterilised container.

Stored seeds must be sampled periodically to check whether there has been any decline in water content and vigour and viability. If water content has remained essentially what it was when the seeds were placed in hydrated storage, and there is no apparent fungal (or bacterial) proliferation, but viability has declined, then the end of the useful storage period will have been reached. Similarly, if visible signs of germination of many of the seeds are apparent, the end of the useful storage period will have been reached. A decline in viability of seeds that have not lost water to any marked extent, or root protrusion by most of the seeds, gives a measure of the time for which hydrated storage is possible under the specific temperature regime used.

Contingencies

Loss of water from seeds indicates that high RH was not maintained, probably because the storage containers were not properly sealed. This leaves uncertain results for the sample, which should be discarded. Loss of viability of seeds during storage may also be the result of maintenance at inappropriate temperatures. This parameter needs to be resolved by trials testing seed responses to a range of temperatures. Seeds may have lost viability because they were originally of poor quality, or of being too immature at harvest.

In cases where there is a high incidence of internally-contaminated seeds in an accession, the contaminants should be isolated and identified, with a view to developing effective means to eliminate them from future collections. Identification of fungi, certainly to genus level, could assist in selection of fungicides that may be more efficacious in combination ('cocktails'), specifically targeting those fungi. Sometimes viruses are present in the seeds, which cannot be eliminated by any treatments. If they can cause serious diseases, the plants must be discarded as soon as viral symptoms are observed.

Contamination may prove to be intractable to any remedial treatment, in which case seeds cannot be stored in this manner, and alternate forms of conservation of the genetic resources must be sought. In such cases, seeds should be set to germinate, with seedlings developed from any seeds that are uninfected being maintained under slow-growth conditions, and/or used to provide alternative explants for *ex situ* conservation, for instance transferred and planted in field genebanks, or other gardens, as appropriate.

SELECTED REFERENCES

Calistru, C., McLean, M., Pammenter, N.W. & Berjak, P. 2000. The effects of mycofloral infection on the viability and ultrastructure of wet-stored recalcitrant seeds of *Avicennia marina* (Forssk.) Vierh. *Seed Science Research*, 10: 341–353.

Pasquini, S., Braidot, S., Petrussa, E. & Vianello, A. 2011. Effect of different storage conditions in recalcitrant seeds of holm oak (*Quercus ilex* L.) during germination. *Seed Science and Technology*, 39: 165–177.

Pritchard, H.W., Wood, C.B., Hodges, S., & Vautier, H.J. 2004. 100-seed test for desiccation tolerance and germination: a case study on eight tropical palm species. *Seed Science and Technology*, 32: 393–403.

Sacandé, M., JØker D., Dulloo, M.E. & Thomsen, K.A., eds. 2004. *Comparative storage biology of tropical tree seeds*. Rome, IPGRI.

Sutherland, J.R., Diekmann, M. & Berjak, P., eds. 2002. *Forest tree seed health*. IPGRI Technical Bulletin No. 6. Rome, IPGRI.

6.4 Standards for *in vitro* culture and slow growth storage

Standards

6.4.1 Identification of optimal storage conditions for *in vitro* cultures must be determined according to the species.

6.4.2 Material for *in vitro* conservation should be maintained as whole plantlets or shoots, or storage organs for species where these are naturally formed.

6.4.3 A regular monitoring system for checking the quality of the *in vitro* culture in slow-growth storage, and possible contamination, should be in place.

Context

In vitro conservation is used for maintenance of plant organs or plantlets in a medium-term time frame (some months up to some years) under non-injurious, growth-limiting conditions. Generally, it is not desirable for long-term conservation (Engelmann, 2011). *In vitro* conservation is preferentially applied to clonal crop germplasm as it also supports safe germplasm transfers under regulated phytosanitary control. Technical documents provide detailed information on the possibilities offered by *in vitro* storage, on the main parameters to consider and on the links and complementarity with other storage technologies, such as field genebanks (Reed *et al.*, 2004; Engelmann, 1999a).

In vitro cultures serve as sources of disease-free materials for distribution, multiplication and a source of explants for cryo-preservation. Safe removal and disposal of infected materials is essential, as it ensures that a pathogen or pest is not released into the environment. Permanent regular monitoring is necessary to avoid accumulation of contamination that might take place during transfers, be transmitted through the air from vessel to vessel or actively transported by vectors like mites and thrips. Breakdown by hyperhydration is another danger, which usually starts in some vessels a little earlier so that a chance exists to rescue the other material when noticed early enough.

Technical aspects

Optimal conditions for slow growth need to be identified prior to storage. These may be achieved by manipulating variables, including light-regime, temperature, and medium composition, individually, or in combination (Engelmann, 1991), but experimentation is generally required to achieve optimal results.

The type and physiological condition of explants is basic to success or failure of *in vitro* slow growth. *In vitro* culture is also used as a preparatory phase to cryopreservation as well as for recovery phases after cryopreservation. Thus, suitable media and conditions for *in vitro* growth of explants need to be developed as a first step. This involves appropriate surface disinfection procedures and germination medium (starting with a standard medium [Murashige and Skoog, 1962], which may need refining). The basal medium may be determined from the literature concerning culture of similar species. Standard protocols have been published and can be used for guidance (including George 1993; Hartmann *et al.*, 2002; Chandel *et al.*, 1995) but in many cases, detailed trials using explants media and growth conditions are critical and custom-made protocols using explants media and growth conditions are needed even if the species are closely related.

Ensuring that the materials are maintained as whole plantlets or shoots, can avoid hyper-hydricity ('vitrification'). For explants of species that naturally grow slowly, no manipulation of media or of culture conditions may be necessary.

Experimentation with a range of permutations and combinations of the means to achieve satisfactory slow growth are imperative when first working with explants of a species. For example, very variable responses to manipulations for slow growth have been recorded for different species of single genera. Maintenance of long-term genetic stability of material stored under slow-growth conditions is imperative (Engelmann, 2011). Optimal storage temperatures for cold-tolerant species may be from 0 to 5 °C or somewhat higher; for material of tropical provenance the lowest temperatures tolerated may be in the range from 15 to 20 °C, depending on the species (Normah *et al.*, 2011; ProMusa; Engelmann, 1999a; Engelmann, 1991).

Various modifications are generally made to culture media, especially reduced levels of minerals, reduction of sucrose content and/or manipulation of the type and concentration of growth regulators, while inclusion of osmotically-active substances (e.g. mannitol) may also be effective (Engelmann, 2011; Engelmann, 1999a). Activated charcoal in the medium may adsorb exuded polyphenolics (Engelmann, 1991).

The type, volume, means of closure and atmosphere in culture vessels constitute important parameters (Engelmann, 2011; Engelmann, 1991), which can be established only by experimentation when working with new material.

Although slow-growth storage is traditionally used for material cultured *in vitro*, plantlets may also be maintained *ex vitro* under growth-restricting conditions. Seedling slow growth in shaded, light-limiting conditions under natural canopies is an inexpensive alternative (Chin, 1996). Furthermore, induction of storage organs *in vitro* can be used for effectively enhancing the conservation period in natural storage organ forming crops (e. g. ginger [Engels *et al.*, 2011], taro, yam, potato etc.).

Contingencies

In vitro culture of explants of woody species may pose particular problems, especially regarding exudation of polyphenolics (Engelmann, 1999b). Associated problems include poor rooting and the explants becoming hyperhydric. Hyperhydration and leaf necrosis developed during slow growth can lead to deterioration of quality and in some cases death of the whole propagules.

In some material, accumulation of covert bacteria may become a gradually increasing obstacle for prolonged slow-growth storage. It may be counteracted by temporary removal of vitamins from the medium or addition of antibiotics, but rarely these measures will be of permanent success. Thus, it may be necessary to discard these cultures from the storage (Abreu-Tarazi *et al.*, 2010; Leifert and Cassels, 2001; Senula and Keller, 2011; Van den Houwe and Swennen, 2000; Van den Houwe *et al.*, 1998).

Within a genepool, there may be large differences in the response to *in vitro* storage between species/varieties, some responding well while others cannot be conserved using this technology, thus making its application impossible (e.g. for coffee [Dussert *et al.*, 1997]). In some species (e.g. yam), storage organs may be formed *in vitro*, but their germination is difficult to attain. This is true also for *in vitro*-derived bulblets in some accessions of a species (e.g. garlic [Keller, 2005]).

In some species, intrinsic genetic instability (e.g. sugar cane) may be enhanced by *in vitro* culture techniques, whereas in others (e.g. cassava) stability over extended storage periods has been demonstrated (IPGRI/CIAT 1994). In the latter cases, somaclonal variation may occur in higher frequencies. In most cases, somaclonal variation is minimized by consequent use of techniques that avoid induction of adventitious shoots or any formation basal callus after cutting. Where callus has formed this needs to be cut off during transfer to the next culture period. To avoid confusion about the reasons for any genetic deviations occurring, thorough observation of uniformity of source explants is needed and chimerism should also be excluded from the donor material (or carefully maintained if needed in variegated plants). Since regular screening by means of molecular markers seems to be too expensive, regular sampling may be undertaken in cases where somaclonal variation is expected to occur.

Dormancy of organs may become a problem, when shoots stop developing (often occurring in species that form *in vitro* storage organs). Additional cutting or application of cytokinins may break dormancy. If this is not successful, then waiting for some time until spontaneous sprouting may be the only (even though uncertain) solution.

SELECTED REFERENCES

Abreu-Tarazi, M.F., Navarrete, A.A., Andreote, F.D., Almeida, C.V., Tsai, S.M. & Almeida, M. 2010. Endophytic bacteria in long-term *in vitro* cultivated "axenic" pineapple microplants revealed by PCRDGGE. *World J. Microbiol Biotechnol.* 26: 555-560.

Benson, E.E., Harding, K. & Johnston, J.W. 2007. Cryopreservation of shoot-tips and meristems. *In* J.G. Day & G. Stacey, eds. *Methods in molecular biology, Vol. 368. Cryopreservation and freeze drying protocols.* 2nd edition, pp. 163–184. Totowa, New Jersey, USA, Humana Press.

Chandel, K.P.S., Chaudhury, R., Radhamani, J. & Malik, S.K. 1995. Desiccation and freezing sensitivity in recalcitrant seeds of tea, cocoa and jackfruit. *Annals of Botany*, 76: 443-450.

Chin, H.F. 1996. Strategies for conservation of recalcitrant species. *In* M.N. Normah, M.K. Narimah & M.M. Clyde, eds. In vitro *conservation of plant genetic resources*, pp. 203-215. Kuala Lumpur.

Dussert, S., Chabrillange, N., Anthony, F., Engelmann, F., Recalt, C. & Hamon, S. 1997. Variability in storage response within a coffee (*Coffea* spp.) core collection under slow growth conditions. *Plant Cell Reports*, 16: 344-348.

Engelmann, F. 1991. *In vitro* conservation of tropical plant germplasm – a review. *Euphytica*, 57: 227–243.

Engelmann, F. ed. 1999a. *Management of field and* in vitro *germplasm collections.* Proceedings of a consultation meeting, 15–20 January 1996. Cali, Colombia, CIAT, and Rome, IPGRI.

Engelmann, F. 1999b. Alternative methods for the storage of recalcitrant seeds – an update. *In* M. Marzalina, K.C. Khoo, N. Jayanthi, F.Y.M. Tsan & B. Krishnapillay, eds. *Recalcitrant seeds*, pp. 159–170. Kuala Lumpur, FRIM.

Engelmann, F. 2011.Use of biotechnologies for the conservation of plant biodiversity. *In vitro Cellular and Developmental Biology – Plant*, 47: 5–16.

Engels, J.M.M., Dempewolf H. & Henson-Apollonio V. 2011. Ethical considerations in agro-biodiversity research, collecting, and use. *J. Agric. Environ. Ethics*, 24: 107–126.

George, E.F. 1993. *Plant propagation by tissue culture. Part 1: The technology.* 2nd edition. Whitchurch, UK, Exegenics.

Hartmann, H.T., Kesler, D.E., Davies, F.T. & Geneve, R.L. 2002. *Plant propagation – principles and practices.* 7th edition. Upper Saddle River, New Jersey, USA, Prentice Hall.

IPGRI/CIAT (IPGRI/International Center for Tropical Agriculture). 1994. *Establishment and operation of a pilot* in vitro *active genebank.* Report of a CIAT-IBPGR collaborative project using cassava (*Manihot esculenta* Crants) as a model. Cali, Colombia, IPGRI and CIAT.

Keller, E.R.J. 2005. Improvement of cryopreservation results in garlic using low temperature preculture and high-quality *in vitro* plantlets. *Cryo-Letters*, 26: 357–366.

Leifert, C. & Cassells, A.C. 2001. Microbial hazards in plant tissue and cell cultures. In Vitro *Cell Dev. Biol. Plant*, 37: 133–138.

Murashige, T. & Skoog, F. 1962. A revised medium for rapid growth and bioassays with tobacco tissue culture. *Physiologia Plantarum*, 15: 473–497.

Normah, M.N., Kean, C.W., Vun, Y.L. & Mohamed-Hussein, Z.A. 2011. *In vitro* conservation of Malaysian biodiversity – achievements, challenges and future directions. *In vitro Cellular and Developmental Biology – Plant*, 47: 26–36.

ProMusa. [Website] (available at: http://www.promusa.org/tiki-custom_home.php).

Reed B.M., Engelmann, F., Dulloo, E. & Engels, J.M.M., eds. 2004. *Technical guidelines for the management of field and in vitro germplasm collections.* Rome, IPGRI/FAO/SGRP-CGIAR.

Senula, A. & Keller, E.R.J. 2011. Cryopreservation of mint – routine application in a genebank, experience and problems. *Acta Hort.*, 908: 467–475.

Van den Houwe, I. & Swennen, R. 2000. Characterization and control of bacterial contaminants in *in vitro* cultures of banana (*Musa* spp.). Acta Hort., 530: 69–79.

Van den Houwe, I., Guns, J. & Swennen, R. 1998. Bacterial contamination in Musa shoot tip cultures. *Acta Hort.*, 490: 485–492.

6.5   Standards for cryopreservation

Standards

6.5.1   The explants selected for cryopreservation should be of highest possible quality, and allow onward development after excision and cryopreservation.

6.5.2   Each step in the cryo-protocol should be tested individually and optimized in terms of vigour and viability in retention of explants.

6.5.3   Means should be developed to counteract damaging effects of reactive oxygen species (ROS) at excision and all subsequent manipulations.

6.5.4   Following retrieval, explants should be disinfected using standard sterile procedures.

Context

Cryopreservation permits cells or tissue to be stored for an indefinite period in LN (–196 °C) where metabolic activities are suspended. Four steps are essential in any cryopreservation protocol: (i) selection, (ii) preculture[1], (iii) cryopreservation techniques, (iv) retrieval from storage, and, (v) seedling or plantlet establishment.

Cryo-protocols should be developed to prevent cryopreservation damages, and could include possible cryo-protection, partial drying, cooling, storage at cryogenic temperatures, rewarming and rehydration. There are two main types of cryopreservation procedures:

---

1   treatment for slow explants' acclimatization to dehydration/cold/freezing

conventional slow freezing, based upon freeze-induced dehydration; and, flash-freezing (vitrification), which involves dehydration prior to cooling (Engelmann, 2011a).

Technical aspects

Selection of explants

Dehydration rate, and how evenly cells and tissue dry, depend on size, and since the vast majority of recalcitrant seeds are too large to be dried rapidly and evenly, they cannot be cryopreserved intact. In addition, cells with water contents ≥ 1.0 g g^{-1}, cannot survive after exposure to cryogenic conditions. Excision and cultivation of suitable explants should be developed specifically for the purpose of cryopreservation. The explants should be as small as possible, but big enough to allow onward development after excision, and after cryopreservation. Higher cellular/tissue uniformity within the explant elevates a chance of cryoprotection of all (or majority) explant cells and its regeneration capacity without callus proliferation. Explants for cryopreservation can be produced from embryonic axes, shoot tips, meristematic and embryogenic tissues. For recalcitrant seeds, excised embryos/axes are the explants of choice for cryopreservation. In the event they are too large, do not withstand the required degree of dehydration, are sensitive to all common modes of surface disinfection, and/or are intractable to culture conditions, explants such as shoot apical meristems are a better option.

For vegetatively propagated species, explants of choice are buds, shoot tips, meristematic, and embryogenic tissue. Not all types of explants are amenable to similar cryoprotection procedures, even when the parent species are relatively closely related taxonomically (Sershen *et al.*, 2007), and responses to cryoprotection procedures need to be ascertained per species as well as genotypes. Under-developed material is generally more susceptible to excision damage; likewise, seeds that have developed/germinated to the stage of visible protrusion of radicals (or other parts of the embryo) should not be selected (Goveia *et al.*, 2004).

Whole anthers or isolated pollen grains can be used for cryopreservation as well. They represent the inherited genetic diversity like seeds, but bearing the male germ units, they usually have the haploid chromosome set only (see Ganeshan, 2008; Rajashekaran, 1994; Weatherhead *et al.*, 1978, for a review). When pollen is conserved, it needs to be embedded in gelatine capsules or paper pouches or packed on a paper strip, with some species requiring dehydration of the pollen prior to storage.

To retrieve the material, anthers or pollens are shed from capsules, pouches or strips at room temperature. The assessment of pollen germination is best undertaken in a germination medium. Viability may be tested by pollen staining, and the results are correlated to pollen germination, although the germination rate will almost always be lower. When the behaviour of a species is still not known, test pollinations are needed to confirm successful fertilization by seed set (Ganeshan *et al.*, 2008; Rajashekaran *et al.*, 1994; Weatherhead *et al.*, 1978).

Probabilistic tools are available which facilitate calculation of the number of propagules to store and retrieve, depending on the objectives, survival after cryostorage, and other parameters (Dussert *et al.*, 2003).

Cryopreservation techniques
It is important that a drying time-course of excised embryos/axes be conducted to identify the drying time required to reduce material to appropriate water content. An additional drying time-course needs to be done after any pre-growth or cryoprotectant treatment.

Cooling rate to LN temperatures is important and needs to be considered in relation to the explant water content. The cryo-protocol should be selected to ensure the water content lies within the range that prevents intracellular ice-crystal formation on cooling and warming, but also avoids desiccation damage to subcellular structure. At the higher end of the water content range to which axes are dried, the faster the cooling rate the better, as very rapid cooling of small specimens tends to be even and minimizes the duration in the temperature range that would permit ice crystallization. The embryos/axes, generally constitute only an insignificant fraction of seed mass and volume, and are suitable for flash-drying, thus overcoming the problem of metabolism-linked damage. On the other hand, cooling rate is less critical for recalcitrant axes flash-dried (using evaporative dehydration) near to their lower limits of tolerance.

Techniques based on dehydration during controlled-rate cooling, have an application when the material to be cryopreserved consists of embryogenic cultures and of shoot tips from temperate species (Engelmann, 2011a). For vegetative material, many protocols and examples of cryopreservation of a range of explants across species using one or more of the procedures, are documented (Benson *et al.*, 2007). In addition, there is a vast number of publications on cryopreservation of apices, other meristematic tissues, embryogenic tissues and dormant buds, and the journal, CryoLetters, is a good source for many of these. Once a successful protocol has been developed for a species, periodic testing of samples extracted from cryopreservation should be carried out, initially after a short storage interval.

Most plant vitrification protocols use cryoprotectants (usually a mixture of penetrating and non-penetrating types). Evaporative dehydration has generally been employed for zygotic embryos/embryonic axes. Although originally developed for apices and somatic embryos, encapsulation-dehydration and the procedure termed vitrification (employing various plant vitrification solutions [PVS]), have also been used in procedures to cryopreserve seed-derived embryos and embryonic axes. A recent overview (Engelmann, 2011b) provides the information that all vitrification protocols developed for somatic embryos, utilizing PVS2. Vitrification using PVS2 has also been used for cryopreservation of shoot tips of a wide range of species from both tropical and temperate provenances, the former including several recalcitrant-seeded and vegetatively propagated species. Another common vitrification solution is PVS3 (Nishizawa *et al.*, 1993) which does not use dimethyl sulphoxide (DMSO) and can therefore be preferred for species that are damaged by DMSO. A range of alternative loading and vitrification solutions have been developed recently, which can be efficiently used for cryopreserving materials which prove sensitive to PVS2 and PVS3 (Kim *et al.*, 2009a; Kim *et al.*, 2009b).

At the lower limits of dehydration tolerated by recalcitrant embryos and axes, generally a proportion of freezable water is retained. During both slow cooling and rewarming, ice crystallization can occur in the freezable water fraction between about -40 and -80 °C. Rewarming at ~37 to 40 °C prevents this, noting that transfer from cryogenic temperatures must be very rapid.

The main cryopreservation techniques and their crucial required parameters are given below:

- controlled-rate cooling: choosing of cryoprotectant (rarely mixture of cryoprotectants); selection of cooling rate (for avoiding crystallization inside of cells);
- encapsulation-dehydration: determination of osmotic dehydration time and it rate of treatment, determination of air desiccation time;
- vitrification: determination of kind of vitrification solution and it time of treatment (assessment of their toxicity); PVS2 should be used on ice.
- droplet freezing: determination of kind of vitrification solution and it time of treatment (assessment of their toxicity).

Retrieval from cryostorage

Rewarming of vitrified germplasm is often undertaken in two steps, the first is slow to allow for glass relaxation, usually at ambient room temperatures. This is followed by more rapid rewarming at ca. 45 °C to avoid ice nucleation (Benson *et al.*, 2011).

Specimens processed by encapsulation-dehydration[2] may be transferred directly onto recovery/germination medium for rapid rewarming, or the cryotubes containing the alginate beads may be placed in a water-bath at 40 °C for 2-3 min. Alternatively, the beads may be rehydrated by transferring them for ~10 min in liquid medium. The removal of the capsule has also been shown to be advantageous (Engelmann *et al.*, 2008). Encapsulation-dehydration has proved to be consistent and successful for shoot tips of many species (González and Engelmann, 2006), somatic embryos of conifers (Engelmann, 2011b), a range of citrus species and varieties, and temperate fruit species (Damiano *et al.*, 2003; Damiano *et al.*, 2007).

To restore metabolic activity in the cell upon rewarming, toxic cryoprotectants must be removed from the cell and the normal water balance gradually restored as the cell is returned to a normal functioning temperature. The original composition of the recovery medium may have to be slightly modified after explants have been dehydrated or cryogen-exposed. With the use of plant vitrification solutions (PVS), after rapid rewarming, a dilution or unloading step (removal of toxic PVS) is necessary (Sakai *et al.*, 2008; Kim *et al.*, 2004).

All steps in cryopreservation could compromise survival, and particularly, warming and rehydration can be accompanied by a burst of reactive oxygen species (ROS)[3] (Whitaker *et al*,. 2010; Berjak *et al.*, 2011). Rewarming and rehydration media should ideally also counteract the deleterious effects of ROS, but it is imperative that means are established to reduce the bursts of (ROS) accompanying excision (Whitaker *et al.*, 2010; Berjak *et a.l*, 2011; Engelmann, 2011a; Goveia *et al.*, 2004). Treatments with cathodic water (an electrolysed dilute solution of calcium chloride and magnesium chloride) had potent anti-oxidative properties, which counteracted the effects of ROS bursts at all stages of a cryopreservation protocol for recalcitrant embryonic axes of *Strychnos gerrardii*, and promoted shoot develo ment (Berjak *et al.*, 2011). The beneficial effects of the treatment are more marked when development of embryos/axes progress during a hydrated storage period, indicating the importance of developmental status of the seeds. It appears that treatment of axes with the non-toxic anti-oxidant, cathodic water, offers both an explanation for previous failures of axes to produce shoots, and an ameliorative treatment to counteract stress-related ROS bursts. Furthermore, the instruments used for embryo/axis

---

2   Encapsulation-dehydration entails the explants being encapsulated in alginate beads and cultured (pre-grown) in a sucrose-enriched liquid medium for periods up to 7 d. Following this they are subjected to dehydration, using a laminar air-flow or flash-drying, or by being exposed to activated silica gel, to dry explants to a water content ~0.25 g g^{-1} (20 percent wmb), and finally cooled rapidly.

3   ROS are highly reactive molecules, often free radicals, which damage proteins, lipids and nucleic acids.

excision can exacerbate ROS production. In this regard, use of a hypodermic needle is likely to cause less trauma than will a surgical blade (Benson *et al.*, 2007). The use of DMSO, a hydroxyl radical scavenger, as a preculture step (before complete severing of the cotyledonary remnants) and as a treatment after their removal, has shown to facilitate shoot development. Other antioxidant substances are also used to counteract ROS formation, e.g. ascorbic acid and tocopherol (Chua and Normah, 2011; Johnston *et al.*, 2007; Uchendu *et al.*, 2010). Survival of plant material can also be assessed based on the enzymatic activity of living plant cells (Mikula, 2006).

Seedling and plantlet establishment

Once excised embryos/embryonic axes have been rewarmed, the next step is to generate and establish a seedling or plantlet to complete the regeneration cycle. Seedling and plantlet establishment requires two steps: (i) its establishment *in vitro* and (ii) establishment *ex vitro* and hardening-off or acclimation. The material recovered from cryostorage, must be introduced to recovery medium initially in the dark. For introduction into *in vitro* culture, the surface of explants require to be disinfected and handled with sterilised instruments, with all procedures being carried out in a laminar air-flow. In conditions where no laminar flow box (clean bench) is available, it may be possible to perform the work in closed clean rooms with thorough room and air disinfection. Embryos and embryonic axes need to be rehydrated for 30 min at ambient temperature in the dark. Where they are directly exposed to a rewarming medium, rehydration should be in a solution of the same composition. Resultant seedlings each producing both a root and a shoot are a measure of successful axis cryopreservation. For vegetatively propagated material, cryostorage is considered successful when shoots are obtained, which can be either rooted or further micropropagated.

After a precautionary culture period in the dark (Touchell and Walters, 2000), explants are usually exposed to conventional growth room lighting conditions and temperature regimes which should be established at the outset as suiting the species and its provenance. Light regimes and temperature for *in vitro* germination and seedling/plantlet development are parameters that may need to be fine-tuned, and transferring explants through several culture phases may be necessary. It is critical that the seedlings and plantlets produced *in vitro* are initially maintained under high RH, which is gradually reduced.

The establishment *ex vitro* and its hardening-off essentially involves transfer of the seedlings/plantlet from slow growth culture or cryopreservation of vegetative material from the heterotrophic *in vitro* condition to a sterile soil-based medium in which the autotrophic condition will develop. Recovery media must contain macro- and micro-nutrients, essential minerals and a carbon source, but may also require addition of growth regulators. Media must have been autoclaved during preparation, and any heat-labile components (if required) filter-sterilised and added subsequently. Suitable germination media for embryos/axes of a variety of species are based on MS (Murashige and Skoog, 1962): however, the MS nutrient medium may be utilized at full-strength, or half- or quarter-strength, as indicated by explant responses when first working with seeds of particular species. Depending on the objective sought, explants recovered from cryopreservation are directly grown into a seedling/plantlet for acclimatization, or a multiplication phase can occur before acclimatization, thus offering the possibility of producing the desired number of copies of the retrieved accession.

Contingencies

It should be noted that protocol development can require more than a single collection and may spread over two or more years due to the seasonal nature of seed availability.

It should be noted that material can be conserved either in LN or above LN in the vapour phase. Storage in the vapour phase is much more expensive and less safe than storage directly in LN. Even if some microbes are suspended in LN, there is not essentially the consequence that they would contaminate the samples, because they pass some washing procedures under sterile conditions upon rewarming. Even if spores may adhere to the surface of the explants, microbes cannot enter them in LN because all such processes are stopped at such low temperatures.

Excised axes may not germinate because of their maturity status. Hence, the collected propagule needs to be placed in hydrated storage and sampled periodically for germination and for performance of excised axes. In the event that neither seed/propagule nor excised embryos/axes geminate, the material may be dead, or dormant. Performing a tetrazolium test will determine whether or not the seeds are viable. If so, then dormancy may be assumed, and investigations to break the dormant condition need to be undertaken.

In the case of most recalcitrant-seeded species, regeneration as practised for orthodox-seeded species is not an option. If there is an unacceptable decline in quality of cryo-stored embryos/axes, the only option would be re-sampling of seeds from the parent population(s) and refining of the procedures. In cases where embryos/embryonic axes continue to be intractable to cryopreservation, then attention must

be focused on the development of suitable alternative explants, ideally derived from seedling/plantlets established *in vitro*.

Material from prolonged *in vitro* culture or *in vitro* storage may no longer be suitable for extracting shoot tips for cryopreservation, since this material may contain or may have accumulated covert bacteria (endophytes) which will break out during recovery from cryopreservation and, thus, hamper cryopreservation entirely. There are instances where explants (e.g. nodal segments) of source material from long-term *in vitro*-maintained cultures are excessively hydrated. In such cases, source material should be cultured *de novo*.

Cultures that have become infected need to be immediately removed from the growth room and destroyed. The most devastating contingency in any growth room is infestation by mites. After removing any cultures showing 'mite tracks', rapid response by disinfesting the facility is required. This is followed by inspection of each culture vessel and removal and destruction of any left which show evidence of mites (which bite through Parafilm™, and spread fungal spores from any infected culture to others).

Depletion of LN in a cryostorage vat or LN freezer would lead to irretrievable loss of all samples. If not detected, electrical or other failure of the temperature control system in a growth room could cause overheating with consequent loss of *in vitro* material.

SELECTED REFERENCES

Benson, E.E. & Bremner, D. 2004. Oxidative stress in the frozen plant: a free radical point of view. *In* B.J. Fuller, N. Lane & E.E. Benson, eds. *Life in the frozen state*, pp. 205-241. Boca Raton, Florida, USA, CRC Press.

Benson, E.E., Harding, K. & Johnston, J.W. 2007. Cryopreservation of shoot-tips and meristems. *In* J.G. Day & G. Stacey, eds. *Methods in molecular biology* vol. 368. *Cryopreservation and freeze drying protocols.* 2nd edition, pp. 163-184. Totowa, New Jersey, USA, Humana Press.

Benson, E.E., Harding, K., Debouck, D., Dumet, D., Escobar, R., Mafla, G., Panis, B., Panta, A., Tay, D., Van den Houwe, I. & Roux, N. 2011. *Refinement and standardization of storage procedures for clonal crops - Global Public Goods Phase 2: Part I. Project landscape and general status of clonal crop* in vitro *conservation technologies.* Rome, SGRP-CGIAR.

Berjak, P., Sershen, Varghese, B. & Pammenter, N.W. 2011. Cathodic amelioration of the adverse effects of oxidative stress accompanying procedures necessary for cryopreservation of embryonic axes of recalcitrant-seeded species. *Seed Science Research*, 21: 187-203.

Chua, S.P. & Normah, M.N. 2011. Effect of preculture, PVS2, and vitamin C on survival of recalcitrant Nephelium ramboutan Ake shoot tips after cryopreservation by vitrification. *Cryo Letters*, 32: 596-515.

Damiano, C., Arias Padró, M. D. & Frattarelli, A. 2007 Cryopreservation of some Mediterranean small fruit plants. *Acta Horticulturae*, 760: 187-194

Damiano, C., Frattarelli, A., Shatnawi, M.A., Wu, Y., Forni, C. & Engelmann, F. 2003. Cryopreservation of temperate fruit species: quality of plant materials and methodologies for gene bank creation. *Acta Horticulturae*, 623: 193-200.

Dussert, S., Engelmann, F. & Noirot, M. 2003. Development of probabilistic tools to assist in the establishment and management of cryopreserved plant germplasm collections. *CryoLetters*, 24: 149-160.

Engelmann, F. 2011a. Germplasm collection, storage and preservation. *In* A. Altman & P.M. Hazegawa, eds. *Plant biotechnology and agriculture – prospects for the 21st century*, pp. 255-268. Oxford, UK, Academic Press.

Engelmann, F. 2011b. Cryopreservation of embryos: an overview. *In* T.A. Thorpe & E.C. Yeung, eds. *Plant embryo culture methods and protocols. Methods in molecular biology*, Vol. 710, Springer Science+Business Media, LLC.

Engelmann, F., González-Arnao, M.T., Wu, Y., & Escobar, R. 2008. The development of encapsulation dehydration. *In* B.M. Reed, ed. *Plant cryopreservation. A practical guide*, pp. 59-75. New York, USA, Springer.

Ganeshan, S., Rajasekharan, P.E., Shashikumar, S. & Decruze, W. 2008. Cryopreservation of pollen. *In* B.M. Reed, ed. *Plant cryopreservation. A practical guide.* pp. 443-464. New York, USA, Springer.

González Arnao, M.T. & Engelmann, F. 2006. Cryopreservation of plant germplasm using the encapsulation-dehydration technique: review and case study on sugarcane. *Cryo Letters*, 27: 155-168.

Goveia, M., Kioko, J.I. & Berjak, P. 2004. Developmental status is a critical factor in the selection of excised recalcitrant axes as explants for cryopreservation: A study of *Trichilia dregeana* Sond. *Seed Science Research*, 14: 241-248.

Johnston, J., W. Harding, K. & Benson, E.E. 2007. Antioxidant status and genotypic tolerance of Ribes *in vitro* cultures to cryopreservation. *Plant Sci.*, 172: 524–534.

Kim, H.H., Cho, E.G., Baek, H.J., Kim, C.Y., Keller, E.R.J. & Engelmann, F. 2004. Cryopreservation of garlic shoot tips by vitrification: Effects of dehydration, rewarming, unloading and regrowth conditions. *Cryo Letters*, 25: 59–70.

Kim, H.H., Lee, Y.G., Shin, D.J., Kim, T., Cho, E.G. & Engelmann, F. 2009a. Development of alternative plant vitrification solutions in droplet-vitrification procedures. *Cryo Letters*, 30: 320–334.

Kim, H.H., Lee, Y.G., Ko, H.C., Park, S.U., Gwag, J.G., Cho, E.G. & Engelmann, F. 2009b. Development of alternative loading solutions in droplet-vitrification procedures. *Cryo Letters*, 30: 291–299.

Mikuła, A., Niedzielski, M. & Rybczyński, J.J. 2006. The use of TTC reduction assay for assessment of *Gentiana* spp. cell suspension viability after cryopreservation. *Acta Physiologiae Plantarum*, 28: 315–324.

Murashige, T. & Skoog, F. 1962. A revised medium for rapid growth and bioassays with tobacco tissue culture. *Physiologia Plantarum*, 15: 473–497.

Nishizawa, S., Sakai, A., Amano, Y. & Matsuzawa, T. 1993. Cryopreservation of asparagus (*Asparagus officinalis* L.) embryogenic suspension cells and subsequent plant regeneration by vitrification. *Plant Sci.*, 91: 67–73.

Rajasekharan, P.E., Rao, T.M., Janakiram, T. & Ganeshan, S. 1994. Freeze preservation of gladiolus pollen. *Euphytica*, 80: 105–109.

Reed, B.M., ed. 2008. *Plant cryopreservation. A practical guide*. New York, USA, Springer.

Sakai, A., Hirai, D. & Niino, T. 2008. Development of PVS-based vitrification and encapsulation-vitrification protocols. *In* B.M. Reed, ed. *Plant cryopreservation. A practical guide*, pp. 33–57. New York, USA, Springer.

Sershen, Berjak, P., Pammenter, N.W. & Wesley-Smith, J. 2011. Rate of dehydration, state of subcellular organisation and nature of cryoprotection are critical factors contributing to the variable success of cryopreservation: studies on recalcitrant zygotic embryos of *Haemanthus montanus*. *Protoplasma*, 249(1): 171–86.

Sershen, Pammenter, N.W., Berjak, P. & Wesley-Smith, J. 2007. Cryopreservation of embryonic axes of selected amaryllid species. *Cryo Letters*, 28: 387–399.

Shatnawi, M.A., Engelmann, F., Frattarelli, A. & Damiano, C. 1999. Cryopreservation of apices of *in vitro* plantlets of almond (*Prunus dulcis* Mill.). *CryoLetters*, 20: 13–20.

Touchell, D. & Walters, C. 2000. Recovery of embryos of Zizania palustris following exposure to liquid nitrogen. *Cryo Letters*, 21: 26–270.

Uchendu, E.E., Leonard, S.W., Traber, M.G. & Reed, B.M. 2010. Vitamins C and E improve regrowth and reduce lipid peroxidation of blackberry shoot tips following cryopreservation. *Plant Cell Rep.*, 29: 25–35.

Weatherhead, M.A., Grout, B.W.W. & Henshaw, G.G. 1978. Advantages of storage of potato pollen in liquid nitrogen. *Potato Res.*, 21: 331–334.

Whitaker, C., Beckett, R.P., Minibayeva, F. & Kranner, I. 2010. Production of reactive oxygen species in excised, desiccated and cryopreserved explants of Trichilia dregeana Sond. *South African Journal of Botany*, 76: 112–118.

6.6   Standards for documentation

Standards

6.6.1   Passport data for all accessions should be documented using the FAO/ Bioversity multi-crop passport descriptors. In addition, accession information should also include inventory, orders, distribution and data user feedback.

6.6.2   Management data and information generated in the genebank should be recorded in a suitable database, and characterization and evaluation data (C/E data) should be included when recorded.

6.6.3   Data from 6.6.1. and 6.6.2 should be stored and changes updated in an appropriate database system and international data standards adopted.

Context

Comprehensive information about accessions is essential for genebank management. Passport data is a minimum, but additional information including geographical (GPS coordinates) environmental (overlaid climate and soil maps) data of the collection site and historical information as well as data on characterization and evaluation of the material are all very useful.

Technical aspects

Due to advances in information technology, it is now relatively simple to record, manage and share information about accessions. All genebanks should use compatible data storage and retrieval systems. The FAO/Bioversity multi-crop passport descriptors (Alercia *et al.,* 2012) should be used by all genebanks as it facilitates data exchange.

Characterization and Evaluation data are produced by users. Such data are useful to the genebank in the management of their collections (Filer, 2012) and to facilitate the consecutive use. Genebanks are recommended to request information feedback on these data.

Management data should be as complete as possible to enable an effective handling of the collection. Most management data are only of internal use to the curator and of limited or no value to others, users and/or recipient genebanks. Therefore, management data should be restricted for use of the collection holder only; a set of the accession history, life form and availability can be extracted for public use. Beside the key data for the accession (passport and characterization data) they should contain the following:

- History (date of acquisition, preliminary numbers, date of changing the numbers, taxonomical determination, name of the specialist who determined the material, cultivation of any donor material in field or greenhouse, way of extracting the *in vitro-* and cryo-material from this donor material).
- Type of storage (*in vitro* or cryopreservation, or hydrated storage in the case of recalcitrant seeds).
- Place of the stored material (cultivation rooms, cryo-tank with concrete placement in rack and box).
- Splitting of the accession in several parties (when material is split in sub-clones, several cryopreservation sets, number of stored tubes).
- Safety duplication (duplication date, duplicated in which institution/country, responsible person there, reference to duplication agreement documents).
- Reference to the protocol used for *in vitro* culture and/or cryopreservation.
- Labelling of the culture vessels (colour codes, barcodes). LN-resistant labels are available, which, if necessary, can be wrapped around already frozen cryotubes.

Further advances in biotechnology will allow phenotypic data to be complemented by molecular data. Bar coding of accessions will be helpful in managing the information and the material and reduces the possibility of making mistakes.

A majority of genebanks now have access to computers and the internet. Computer-based systems for storing data and information allow for comprehensive storage of all information associated with the management of *in vitro* and cryopreservation collections. Germplasm information management systems such as GRIN-Global (2011) have specifically been developed for universal genebank documentation and information management. The adoption of data standards which today exist for most aspects of genebank data management helps make the information management easier and improves use and exchange of data. Sharing accession information and making it publicly available for potential germplasm users is important to facilitate and support the use of the collection. Ultimately, conservation and usability of conserved germplasm are promoted through good data and information management

Contingencies

Loss or incomplete documentation reduces the value of an accession, to the point of making it unusable. Inappropriate material (e.g. not LN-resistant labels) can cause loss of data. In large collections, skill of the workers becomes a very important factor. Risks of inadequate data entries must be clearly indicated. In complicated collections, active access to management data should be limited to the responsible persons only.

SELECTED REFERENCES

Alercia, A., Diulgheroff, S. & Mackay, M. 2012. FAO/Bioversity *Multi-Crop Passport Descriptors* (MCPD V.2). Rome, FAO and Bioversity International (available at: http://www.bioversityinternational.org/uploads/tx_news/1526.pdf).

Filer, D.L. 2012. BRAHMS Version 7.0. Department of Plant Sciences, University of Oxford, UK (available on http://herbaria.plants.ox.ac.uk/bol/).

USDA, ARS, Bioversity International, Global Crop Diversity Trust. GRIN-Global. Germplasm Resource Information Network Database - Version 1 (available at: http://www.grin-global.org/index.php/Main_Page).

6.7 Standards for distribution and exchange

Standards

6.7.1 All germplasm should be distributed in compliance with national laws and relevant international conventions.

6.7.2 All samples should be accompanied by a complete set of relevant documents required by the donor and the recipient country.

6.7.3 The supplier and recipient should establish the conditions to transfer the material and should ensure adequate re-establishment of plants from *in vitro*/ cryopreserved material.

Context

Germplasm distribution is the supply of a representative sample from a genebank accession in response to requests from germplasm users. There is a continuous increase in demand for genetic resources to meet the challenges posed by climate change, by changes in virulence spectra of major insect pests and diseases, by invasive alien species and by other end-users needs. This demand has led to wider recognition of the importance of using germplasm from genebanks, which ultimately determines the germplasm distribution. It is important that distribution of germplasm across borders adheres to international norms and standards relating to phytosanitary regulations and according to provisions of relevant international treaties and conventions on biological diversity and plant genetic resources.

Technical aspects

The two international instruments that govern the access of genetic resources are the ITPGRFA and the CBD. The ITPGRFA facilitates access to PGRFA, and provides for the sharing of benefits arising from their utilization. It has established a multilateral system for PGRFA for a pool of 64 food and forage crops (commonly referred to as Annex 1 crops to the Treaty), which are accompanied SMTA for distribution. SMTA can also be used for non-Annex 1 crops; however, other models are also available. Access and benefit-sharing under CBD is according to its Nagoya Protocol. Both the ITPGRFA and CBD emphasize the continuum between conservation and sustainable utilization, along with facilitated access and equitable sharing of benefits arising from use.

All accessions should be accompanied with the required documentation such as phytosanitary certificates and import permits, as well as passport information. The final destination and the latest phytosanitary import requirements for the importing country (in many countries, regulations are changed frequently) should be checked before each shipment. Germplasm transfer should be carefully planned in consultation with the national authorized institute, for the appropriate documentation, such as an official phytosanitary certificate, that complies with the requirements of the importing country. The recipient of the germplasm should provide the supplying genebank with information concerning the documentation required for the importation of plant material, including phytosanitary requirements.

Most recalcitrant-seeded species are long-lived perennials that do not reproduce until they are several years old. Thus, regeneration is not a rapid way to bulk up sample sizes to meet demand. If the sample is in the form of alternate explants *in vitro*, multiplication before the production of independent plantlets is possible, but a request must be made in advance.

Germplasm should reach its destination in good condition and so adverse environmental conditions during transport and clearing customs should be minimized. A reliable courier service having experience in dealing with customs is recommended. The time span between receipt of a request for germplasm and the dispatch of the materials should be kept to a minimum to enhance the efficiency of the genebank function. If the sample is in the cryopreserved state and is being transferred to another genebank where it will continue to be cryopreserved, the sample must be shipped in a LN 'dry shipper'.

If the sample will be set out to grow immediately on receipt, it can be rewarmed, rehydrated and encapsulated in a calcium alginate bead prior to dispatch. Such synthetic seeds were originally developed for somatic embryos, but can successfully maintain in good condition cryopreserved excised embryos/axes that have been rewarmed and rehydrated, for at least 10 days at 16 °C without germination (radicle protrusion) being initiated. Germination and seedling/plantlet establishment of synthetic seeds is possible both *in vitro* and could succeed in sterile seedling mix. It is an option also for other small explants from cryopreservation, but the technique is applied in few cases only.

Plantlets derived from *in vitro* slow-growth storage or cryopreservation should be sent in appropriate containers. Recipients of *in vitro*/cryopreserved material need to have the possibility to transfer the material to pots or to the field, or be able to make such arrangements.

Sterile plastic bags, which may contain special aeration zones, are recommended for sending of *in vitro* plantlets. If glass is used, sufficient stuffing of the containers and declaration of fragility need to be ensured. In cases of glass and plastic vessels, also the right orientation of the containers needs to be indicated.

Contingencies

Poor handling, including inappropriate packaging or delay in shipment, can lead to loss of viability and the loss of material. Thus, it is very important that the supplier and recipient have established the condition under which the material is transferred and that the prerequisite for adequate re-establishment of plants is ensured.

SELECTED REFERENCES

Rao, N.K., Hanson, J., Dulloo, M.E., Ghosh, K., Nowell, D. & Larinde, M. 2006. Germplasm distribution. *In Manual of seed handling in genebanks.* Handbooks for Genebanks No. 8. Rome, Bioversity International.

6.8   Standards for security and safety duplication

Standards

6.8.1   A risk management strategy should be implemented and updated as required that addresses physical and biological risks identified in standards including issues such as fire, floods and power failures.

6.8.2   A genebank should follow the local Occupational Safety and Health requirements and protocols. The cryo-section of a genebank should adhere to all safety precautions associated with using LN.

6.8.3   A genebank employs the requisite staff to fulfil all routine responsibilities to ensure that the genebank can acquire, conserve and distribute germplasm.

6.8.4   A safety duplicate sample of every accession should be stored in a geographically distant genebank under best possible conditions.

6.8.5   The safety duplicate sample should be accompanied by relevant documentation.

Context

It is of the utmost importance that the physical infrastructure of any genebank as well as the safety of its staff be protected so as to ensure that the conserved germplasm is safe from any threatening external factors. To manage a germplasm collection successfully, a genebank also requires skilled and trained staff. Management involves not only the maintenance of the collection and its data but also an assessment of risks from human activity or those naturally caused. There are particular hazards associated with the use of LN.

The physical security of the collections also requires a safe duplication of the collections in a geographically distant location, under the same conditions. In case of natural/physical catastrophe (fire, flood), this duplication might be used to re-build the collections. In addition to the sample itself, safety duplication involves the duplication of information that implies database backup.

Technical aspects

A genebank should implement and promote systematic risk management that addresses the physical and biological risks in the every-day environment. It should have in place a written risk management strategy on actions that need to be taken whenever an emergency occurs in the genebank concerning the germplasm or the related data. This strategy and an accompanying action plan must be regularly reviewed and updated to take account of changing circumstances and new technologies, and well publicized among their genebank staff.

The occupational health and safety of the staff should also be considered. The cryostorage area should be well ventilated with forced air extraction, and oxygen monitors should be in place. Leakage of LN into cryovials is potentially dangerous; therefore, appropriate vessels that are specifically designed for the purpose should be used, and the manufacturers' instructions should be strictly adhered to. To reduce risk of personal injury operators should wear protective clothing, gloves and face masks.

Supplies of LN must always be available, and it is vital that levels of LN are maintained. The cryogenic storage tanks are supposed to be placed in an appropriated location: aerated and with temperature less than 50 °C. Maintenance of the level of LN in storage containers is absolutely critical; if all the LN evaporates, the entire contents of the storage container must be discarded.

For the maintenance of viability of samples, the temperature of the tissue must be kept below the glass transition temperature[1]. Care must be taken that when removing a vial from a cryo-cane or from a cryo-box that the temperature of the remaining vials does not increase to the glass transition temperature. Vials should not be labelled with conventional adhesive labels, as they will come off at LN temperatures. The use of a dedicated PC-operated label printer allows specific cryovial labels to be printed, recording information and a unique barcode. The manufacturer's recommendations about which vial to use for which particular purpose should be adhered to.

Active genebank management requires well-trained staff, and it is crucial to allocate responsibilities to suitably competent employees. A genebank should therefore, have a plan in place for personnel, and a corresponding budget allocated regularly so as to guarantee that a minimum of properly trained personnel is available to fulfil the responsibilities of ensuring that the genebank can acquire, conserve and distribute germplasm. Access to disciplinary and technical specialists in a range of subject areas is desirable. Staff should have adequate training acquired through certified training and/or on-the-job training and training needs should be determined as they arise.

For the physical security of the collections, safe duplication of the collections in a geographically distant location under the same conditions should be considered. In case of natural/physical catastrophe (fire, flood), this duplication might be used to re-build the collections. The duplicating bank should be located somewhere that is politically and geologically stable, and at an elevation that rising sea levels will not be a problem. The storage conditions for the safety duplicate should be as good as those of the initial collection.

---

1   In PVS2, one of the most commonly-used cryo-protectant solutions, glass transitions occur at about -115 °C.

Safety duplication requires a signed legal agreement between depositing and storing or repository genebank. The latter has no entitlement to the use and distribution of the germplasm. The access to the collections should be controlled to avoid unauthorized usage.

Samples for the safety duplicate should be prepared in the same way as the initial collection. It is the responsibility of the depositor to ensure that the safety duplicate is of good quality. To prevent deterioration in transit to the receiving bank, cryopreserved samples should be dispatched in a LN dry shipper, and transit should be as rapid as possible.

Contingencies

When suitably trained staff is not available, or when there are time or other constraints, a solution might be to outsource some of the work or call for assistance from other genebanks.

Unauthorized entry to the genebank facilities can result in direct loss of material, and jeopardize the collections through introduction of pest and diseases.

LN containers are often contaminated with fungi or bacteria. If samples are stored in the liquid phase of the nitrogen, contamination of the sample can occur.

Liability issues may arise if material deteriorates in transit. Therefore, all eventualities need to be adherent to the consignment agreement.

SELECTED REFERENCES

Benson, E.E. 2008. Cryopreservation of phytodiversity: a critical appraisal of theory and practice. *Critical Reviews in Plant Sciences*, 27: 141-219.

Volk, G.M. & Walters, C. 2006. Plant vitrification solution 2 lowers water content and alters freezing behaviour in shoot tips during cryoprotection. *Cryobiology*, 52: 48-61.

Annex 1:
List of Acronyms

AFLP	Amplified fragment length polymorphisms
BRAHMS	Botanical Research and Herbarium Management System
CBD	Convention on Biological Diversity
CGRFA	Commission on Genetic Resources for Food and Agriculture
DMSO	Dimethyl sulphoxide
ELISA	Enzyme-linked immunosorbent assay
ENSCONET	European Native Seed Conservation Network
EST-SSR	Expressed sequence tags - simple sequence repeats
FAO	Food and Agriculture Organization of the United Nations
GBS	Genotyping-by-sequencing
GPS	Global Positioning System
GRIN	Germplasm Resources Information Network
GWS	Genome-wide selection
ICIS	International Crop Information System
IPGRI	International Plant Genetic Resources Institute (now called Bioversity International)
IPPC	International Plant Protection Convention
ISTA	International Seed Testing Association
ITPGRFA	International Treaty on Plant Genetic Resources for Food and Agriculture

LN	Liquid nitrogen
MS	Murashige and Skoog's nutrient medium
MSB Kew	Millennium Seed Bank, Kew Gardens
MTA	Material Transfer Agreement
NaOCl	Sodium hypochlorite
N_e	Effective population size
NPGS	National Plant Germplasm System
OIV	International Organisation for Vine and Wine
OSH	Occupational Safety and Health
PGRFA	Plant Genetic Resources for Food and Agriculture
PVS	plant vitrification solutions
RAPD	Random amplified polymorphic DNA
RFID	Radio- Frequency Identification
RFLP	Restriction fragment length polymorphism
RH	Relative humidity
ROS	Reactive oxygen species
SID	Seed Information Database
SMTA	Standard Material Transfer Agreement
SNP	Single Nucleotide Polymorphism
SOPS	Standard Operating Procedures
SSR	Simple sequence repeats
UPOV	International Union for the Protection of New Varieties of Plants
USDA	United States Department of Agriculture
WTO/SPS	World Trade Organization/ Sanitary and Phytosanitary Agreement

Annex 2:
Glossary

Accession: A distinct, uniquely identifiable sample of seeds representing a cultivar, breeding line or a population, which is maintained in storage for conservation and use.

Accession number: A unique identifier that is assigned by the curator when an accession is entered into a collection. This number should never be assigned to another accession.

Active collection: A set of germplasm accessions that is used for regeneration, multiplication, distribution, characterization and evaluation. Active collections are maintained in short to medium-term storage and usually duplicated in a base collection maintained in medium- to long-term storage.

Barcode: A computerized coding system that uses a printed pattern or bars on labels to identify germplasm accessions. Barcodes are read by optically scanning the printed pattern and using a computer program to decode the pattern.

Base Collection: A set of accessions, each of which should be distinct and, in terms of genetic integrity, as close as possible to the sample provided originally, which is preserved for the long-term future. The base collection for a crop genepool or any species may be dispersed among several institutions - a practice which is likely to increase with the development of crop networks. Normally, seeds will not be distributed from the base collection directly to users.

Characterization: The recording of highly heritable characters that can be easily seen and are expressed in all environments.

Collection: A group of germplasm accessions maintained for a specific purpose under defined conditions.

Cryopreservation or **Cryostorage:** The storage of plant organs in liquid nitrogen (-196 °C) or above, in its vapour phase, (maximum -140 °C). In the genebank context, it is used for buds, shoot tips, other meristematic and embryogenic tissue, explants from recalcitrant and (in special cases) entire orthodox seeds, pollen and somatic embryos. In most cases *in vitro* phases before and/or after the storage phase proper are involved.

Cryopreservation of pollen: Pollen grains are possible targets in some plant families. As in seeds, there are species with "orthodox" pollen and such with "recalcitrant" behaviour. Dehydration of pollen may be necessary before cryopreservation, but pollens of some species are easily storable without previous treatment. For regeneration from stored pollen samples, a crossing partner needs to be present to obtain the finally requested plant material via seed set and germination.

Database: An organized set of interrelated data assembled for a specific purpose and held in one or more storage media.

Descriptor: An identifiable and measurable trait, characteristic or attribute observed in an accession that is used to facilitate data classification, storage, retrieval and use.

Descriptor list: A collection of all individual descriptors of a particular crop or species.

Distribution: The process of supplying samples of germplasm accessions to breeders and other users.

Documentation: The organized collection of records that describe structure, purpose, operation, maintenance, and data requirements.

Donor: An institution or individual responsible for donating germplasm.

Dormancy: The state in which certain live seeds do not germinate, even under normally suitable conditions.

Equilibrium moisture content: The moisture content at which a seed is in equilibrium with the relative humidity of the surrounding air.

Evaluation: The recording of those characteristics whose expression is often influenced by environmental factors.

***Ex situ* conservation:** The conservation of biological diversity outside its natural habitat. In the case of plant genetic resources, this may be in seed genebanks, *in vitro* genebanks or as live collections in field genebanks.

Field: Plot of land with defined boundaries within a place of production on which a commodity is grown.

Genebank: A centre for conserving genetic resources under suitable conditions to prolong their lives.

Genetic diversity: The variety of genetic traits that result in differing characteristics.

Genetic drift: Changes in the genetic composition of a population when the number of individuals is reduced below the frequency of certain alleles within it.

Genotype: The genetic constitution of an individual plant or organism.

Germplasm: The genetic material that forms the physical basis of heredity and that is transmitted from one generation to the next by germ cells.

Germination: The biological process that leads to the development of a seedling from a seed. Radicle emergence is the first visible sign of germination, but may be followed by no further growth or by abnormal development. According to ISTA rules, only seedlings showing normal morphology are considered to have germinated.

Germination testing: A procedure to determine the percentage of seeds that are capable of germinating under a given set of conditions.

In vitro **culture:** The cultivation of plant organs or entire plants on artificial nutrient medium in glass or plastic containers. Using *in vitro* culture of vegetatively propagated crops includes several options, including micropropagation, virus elimination via meristem culture and slow-growth storage. *In vitro* culture is also used as a preparatory phase to cryopreservation as well as for recovery phases after cryopreservation (see also slow-growth storage)

Isotherm: A graph showing the relationship between seed moisture content and percentage relative humidity.

Landrace: A crop cultivar that has evolved through many years of farmer-directed selection and that is specifically adapted to local conditions; landraces are usually genetically heterogeneous.

Long-term conservation: The storage of germplasm for a long period, such as in base collections and duplicate collections. The period of storage before seeds need to be regenerated varies, but is at least several decades and possibly a century or more. Longterm conservation takes place at sub-zero temperatures.

Medium-term conservation: The storage of germplasm in the medium-term such as in active and working collections; it is generally assumed that little loss of viability will occur for approximately ten years. Medium-term conservation takes place at temperatures between 0 °C and 10 °C.

Moisture content (wet-weight basis): The weight of free moisture divided by the weight of water plus dry matter, expressed as a percentage.

Monitoring: The periodic checking of accessions for viability and quantity.

Monitoring interval: The period of storage between monitoring tests.

Most-original-sample (MOS): A sample of seeds that have undergone the lowest number of regenerations since the material was acquired by the genebank, as recommended for storage as a base collection. It may be a subsample of the original seed lot or a seed sample from the first regeneration cycle if the original seed lot required regeneration before storage.

Multiplication: The representative sample of an accession grown to multiply the quantity of conserved material for distribution.

Orthodox seeds: Seeds that can be dried to low moisture content and stored at low temperatures without damage to increase seed longevity.

Pathogen: A living microorganism such as a virus, bacterium or fungus that causes disease in another organism.

Passport data: Basic information about the origin of an accession, such as details recorded at the collecting site, pedigree or other relevant information that assists in the identification of an accession.

Pedigree: The record of the ancestry of a genetic line or variety.

Phenotype: The external appearance of a plant that results from the interaction of its genetic composition (genotype) with the environment.

Phytosanitary: Pertaining to plant quarantine

Phytosanitary certificate: A certificate provided by government plant health personnel to verify that seed material is substantially free from pests and diseases.

Pollination: The process in which pollen is transferred from an anther to a receptive stigma by pollinating agents such as wind, insects, birds, bats, or the opening of the flower itself.

Population: A group of individual plants or animals that share a geographic area or region and have common traits.

Propagule: Any structure with the capacity to give rise to a new plant, whether through sexual or asexual (vegetative) reproduction. This includes seeds, spores, and any part of the vegetative body capable of independent growth if detached from the parent.

Quarantine: The official confinement of introduced germplasm subject to phytosanitary regulations to ensure that it does not carry diseases or pests injurious to the importing country.

Random sample: A sample drawn at random from a larger group.

Recalcitrant seed: Seeds that are not desiccation-tolerant; they do not dry during the later stages of development and are shed at water contents in the range of $0.3 - 4.0$ g g^{-1}. The loss of water rapidly results in decreased vigour and viability, and seed death at relatively high water contents.

Regeneration: Grow-out of a seed accession to obtain a fresh sample with high viability and numerous seeds.

Regeneration standard: The percentage seed viability at or below which the accession must be regenerated to produce fresh seeds.

Relative humidity (RH): A measure of the amount of water present in the air compared to the greatest amount possible for the air to hold at a given temperature, expressed as a percentage. It differs from *absolute humidity*, which is the amount of water vapour present in a unit volume of air, usually expressed in kilograms per cubic meter.

Safety duplication: A duplicate of a base collection stored under similar conditions for long-term conservation, but at a different location to insure against accidental loss of material from the base collection.

Sample: A part of a population used to estimate the characteristics of the whole.

Silica gel: An inert chemical that absorbs water from its surroundings and will give up this water by evaporation when heated.

Seed viability: The capacity of seeds to germinate under favourable conditions.

Slow-growth storage *in vitro*: The maintenance of plant organs or whole plants under conditions that slow down the speed of plant development to reduce the necessary labour input and the frequency of transfers which might be accompanied by risk of infections and stress conditions which eventually would endanger genetic stability. The main method to slow down development is temperature reduction with appropriate temperature being taxon-dependent. At the end of a subculture phase transfer to new medium is necessary, with or without a multiplication step and sometimes warm culture periods for re-establishment.

Sorption isotherm: *See isotherm.*

Storage life: The number of years that a seed can be stored before seed death occurs.

Tetrazolium test: A test for viability in which moist seeds are soaked in a solution of triphenyl tetrazolium chloride.

Trait: a recognizable quality or attribute resulting from interaction of a gene or a group of genes with the environment.

Viability test: A test on a sample of seeds from an accession that is designed to estimate the viability of the entire accession.

Variety: A recognized division of a species, next in rank below subspecies; it is distinguishable by characteristics such as flower colour, leaf colour and size of mature plant. The term is considered synonymous with *cultivar*.

Photo credits

Cover:

Graphic composition adapted from FAO Mediabase and other internet sources

Internal Pages: